# Decentralization and the Social Economics of Development

## Lessons from Kenya

# Decentralization and the Social Economics of Development
## Lessons from Kenya

Edited by

**Christopher B. Barrett**

*Department of Applied Economics and Management, Cornell University, Ithaca, New York*

**Andrew G. Mude**

*International Livestock Research Institute, Nairobi, Kenya*

**John M. Omiti**

*Kenya Institute of Public Policy Research and Analysis (KIPPRA), Nairobi, Kenya*

www.cabi.org

**CABI is a trading name of CAB International**

CABI Head Office
Nosworthy Way
Wallingford
Oxfordshire OX10 8DE
UK

Tel: +44 (0)1491 832111
Fax: +44 (0)1491 833508
E-mail: cabi@cabi.org
Website: www.cabi.org

CABI North American Office
875 Massachusetts Avenue
7th Floor
Cambridge, MA 02139
USA

Tel: +1 617 395 4056
Fax: +1 617 354 6875
E-mail: cabi-nao@cabi.org



A catalogue record for this book is available from the British Library, London, UK.

Library of Congress Cataloging-in-Publication Data

Decentralization and the social economics of development: lessons from Kenya/ edited by Christopher B. Barrett, Andrew G. Mude, John M. Omiti.
p. cm.
Includes bibliographical references and index.

ISBN 978-1-84593-269-5 (alk. paper)

1. Rural development—Kenya. 2. Decentralization in government—Kenya. 3. Kenya—Economic conditions. 4. Kenya—Social conditions. I. Barrett, Christopher B. (Christopher Brendan) II. Mude, Andrew G. III. Omiti, John. IV. Title.

HN793.D43 2007
338.96762--dc22

2007002114

ISBN-13: 978 1 84593 269 5

Typeset by SPi, Pondicherry, India.
Printed and bound in the UK by Cromwell Press, Trowbridge.

# Contents

# Contributors

**David M. Amudavi** *is a Lecturer in the Department of Agricultural Education and Extension, Egerton University, PO Box 536 Njoro, Kenya, and a Post Doctoral Fellow, International Centre of Insect Physiology and Ecology (ICIPE), Nairobi, Kenya.*

**Christopher B. Barrett** *is International Professor in the Department of Applied Economics and Management, Cornell University, 315 Warren Hall, Ithaca, NY 14853, USA. He is also Editor of the American Journal of Agricultural Economics and Co-Director of the African Food Security and Natural Resources Management Program at Cornell University.*

**Godana J. Doyo** *is the District Drought Management Officer for the Arid Lands Resources Management Project in Marsabit, PO Box 155, Marsabit, Kenya.*

**Guyo O. Haro** *is the Project Manager for the GTZ/GEF Indigenous Vegetation Project, PO Box 204, Marsabit, Kenya.*

**Heidi Hogset** *is Associate Professor in the Department of Economics, Informatics and Social Science, Molde University College NO-6402 Molde, Norway.*

**John G. McPeak** *is Assistant Professor in the Department of Public Administration, Syracuse University, 336 Eggers Hall, Syracuse, NY 13244, USA.*

**Andrew G. Mude** *is an Economist at the International Livestock Research Institute, PO Box 30709, Nairobi, Kenya.*

**Godiah L. Mugunieri** *is a Research Scientist at the Kenya Agricultural Research Institute (KARI), PO Box 362, 00902 Kikuyu, Kenya.*

**Kioko Munyao** *is a Food Security and Policy Adviser with World Vision Canada, 1 World Drive, Toronto, Ontario L5T 2Y4, Canada.*

**Elizabeth Nambiro** *is a PhD candidate in the Institute of Plant Nutrition, University of Bonn, Karlrobert-Kreiten-Str. 13, 53115 Bonn, Germany.*

**Isaac K. Ngugi** *is a Research Fellow in the Tegemeo Institute of Agricultural Policy and Development, Egerton University, PO Box 20498, Nairobi, Kenya.*

**James K. Nyoro** *is the Director of Tegemeo Institute of Agricultural Policy and Development, Egerton University, PO Box 20498, Nairobi, Kenya.*

**John M. Omiti** *is a Senior Policy Analyst at the Kenya Institute of Public Policy Research and Analysis (KIPPRA), PO Box 56445, Nairobi, Kenya.*

**Sharon M. Osterloh** *is a Social Science Lecturer in the Social Science Department at Edgewood College, 1000 Edgewood College Drive, Madison, WI 53711, USA.*

# Acknowledgements

The chapters of this volume represent a significant output of the Strategies and Analysis for Growth and Access (SAGA) project, a USAID-funded cooperative agreement directed by Cornell and Clark Atlanta Universities. SAGA brings together policy makers, researchers and other experts with an interest in African development. SAGA's goals are to conduct high-quality, policy-relevant research, and to increase African capacity to produce such research on key issues affecting economic growth and welfare in Africa. While SAGA's geographical coverage spans several countries in western, southern and eastern Africa and its research cuts across issues of health, education, vulnerability and empowerment, this book highlights the insights generated from research carried out by a range of collaborators in Kenya. Specifically, this volume collects research looking into the effect that decentralization, in its various forms, plays in stimulating rural development, empowering individuals and groups, and improving rural well-being. The institutional correlates of successful and unsuccessful decentralization efforts are also studied.

The volume could not have reached fruition were it not for the dedication and contribution of the authors and co-authors of the chapters included herein. We extend our deep gratitude and appreciation for their hard work and commitment to this endeavour. Several other SAGA–Kenya researchers produced rigorous and insightful research around the parallel theme of 'reducing risk in rural Kenya'. They too were valued members of the SAGA–Kenya team. We are indebted to them and sincerely thank and recognize Joseph Karugia, Paswel Marenya, Hezron Nyangito, Rose Nyikal, Walter Odhiambo and Willis Oluoch-Kosura for their efforts.

Early versions of most of these chapters were presented in a conference showcasing SAGA–Kenya outputs, held at the Grand Regency hotel in Nairobi on 10 February 2005. The preliminary chapters benefited greatly from discussions and comments brought up by a distinguished set of participants that included members of the Kenyan parliamentary select committee on agriculture, a broad range of social scientists, development practitioners, representatives of non-governmental agencies and international donor and research organizations. We thank all of the conference participants for their insights and contributions; they have certainly added to the value of this volume. The Institute of Policy Analysis and Research, based in Nairobi, deserves special mention for its role in organizing the conference, as well as generally coordinating SAGA's research efforts in Kenya across participating local institutions.

We are particularly grateful to SAGA for generously supporting most of the research presented in this volume, as well as its publication. In this regard, we especially wish to thank Mesfin Bezuneh, Kelly Doss, Patti Mason, Philip Neuwirth, David Sahn and Diana Wixson of SAGA, as well as Rita Agarwal, Marty Hanratty, Yoon Lee, Tim Mahoney and Borany Penh of USAID. Sue Snyder of Cornell deserves our deep gratitude for the excellent job she did in preparing the final manuscript. Cheryl Mrozowski and Gail Keenan provided excellent administrative support. We thank and acknowledge Elsevier Science for granting us permission to reprint the Haro, Doyo and McPeak paper, which previously appeared in *World Development* (in February 2005: 33(2), 285–299), as Chapter 7. Sincere thanks also go to Nigel Farrar, Meredith Carroll, Rachel Cutts and Mabvuto Hardy of CAB International for all their help with the publication and to Jaya Bharathi of SPi-Publishing for careful typesetting.

Last but certainly not least, we thank our loved ones for their patience and understanding as we took time that could have been spent with them to complete this project. Their love and support make such endeavours worthwhile.

# 1 Decentralization and the Social Economics of Development: An Overview of Concepts and Evidence from Kenya

C.B. BARRETT, A.G. MUDE AND J.M. OMITI

## 1 The Concept and Promise of Decentralization

The last two decades or so have witnessed the steady advance of decentralization – devolved administrative, political and fiscal authority from central government to regional and local jurisdictions – within developing countries (World Bank, 1999). In sub-Saharan Africa, where colonial and post-colonial political dynamics favoured highly centralized systems of government, decentralization has proceeded more slowly and met with more limited success than in other regions (Adamolekun, 1999). Nevertheless, with increasing internal and external pressures to reform the structure of public administration and devolve authority and resources to regional and local level entities, more African governments are expected to travel further along the path of decentralization in coming years (World Bank, 1999; Prud'homme, 2003; Smoke, 2003).

This volume aims to inform that process, especially as it evolves in Kenya and other east African nations, through careful examination of a range of recent, related experiences in rural Kenya. Some of these experiments with decentralization appear to have been reasonably successful in achieving specific development goals – e.g. expanding access to agricultural extension services, helping resolve intercommunity conflicts over natural resources – while others have been far less successful. The variation in the effectiveness of different forms of decentralization in generating desired outcomes provides an opportunity to try to identify some useful common findings that might inform future efforts at devolving decision-making and resource-allocation authority. That is the broad aim of this volume.

Proponents of decentralization often argue that it promotes allocative and productive efficiency in the provision and use of public services, improves governance and political accountability and provides a more effective governing vehicle for advancing pro-poor policies. These arguments in favour of decentralization are grounded in the ideal of a representative government that is responsible for, and accountable to, its constituents and is able to effectively and equitably cater to their needs. Proponents maintain that where the distribution of resource endowments within nations is heterogeneous, and where the needs, constraints and aspirations of the people vary across communities and regions, a centralized system of government suffers informational disadvantages and transactions as well as search costs that render it ill-equipped to identify and provide an effective mix and distribution of services. Moreover, a central government monopoly supplier of public goods and services faces less (typically, no) competition and may be able to demand bribes or other rents that can be extinguished through interjurisdictional competition following decentralization. The core idea is that sub-national and local governments, by virtue of their proximity to the people concerned with policy outcomes, have better access to local information, are more directly accountable to constituents, can more effectively and quickly identify and articulate regional and community-specific needs, and are thus better placed to allocate and extract resources more efficiently than higher-level organs of government.

Because decentralization is perceived to strengthen local institutions and empower communities to manage their resources more effectively, as the larger development agenda has become increasingly pro-poor and focused on 'bottom-up' approaches to community-driven development, this has pushed a wide-ranging programme of decentralization in developing countries. Throughout the 1990s, the development community increasingly associated public sector decentralization reforms with larger poverty alleviation goals by lauding its potential to incubate good governance and improve the legitimacy, accountability and responsiveness of local government (Smoke, 2003). It was argued that devolving administrative authority to regional and local institutions, and facilitating accountability by requiring local representatives to be elected by constituents, would empower the people to engage the local political process and demand effective provision of services. As local governments are most likely to be familiar with local circumstances, they were also perceived to be better equipped to distribute resources more equitably and target poverty more effectively.

Advocacy for decentralization was also politically adept on the part of donors as proponents could voice concern for good governance and put effort into championing it without directly interfering in nations' internal politics. As such, major development agencies and multilateral and bilateral donors weary of corrupt central governments, which hijacked development initiatives, squandered public resources and lacked accountability, began vigorously promoting decentralization as a means to extinguish the culture of corruption and administrative paralysis in central governments. Furthermore, non-governmental organizations (NGOs), community-based organizations (CBOs) and private organizations, seen as the new instruments for pro-poor development, could be funded directly in a decentralized regime. Such sentiment was formalized in the World Bank's Governance and Civil Service Reform Program of the early 1990s which explicitly championed decentralization among client countries (Crook, 2003).

Sceptics of decentralization counter that it facilitates elite capture by local government and the persistence of anachronistic institutions based on patron–client relations, that it indulges a cultural bias towards conservatism because progressive leadership is so scarce, and that it moves allocative decisions further out of the limelight, thereby permitting greater corruption or mismanagement of scarce resources. Another common critique revolves around economies of scale. If decentralization increases transactions and search costs for delivering public goods and services, these effects may trump any purported informational efficiencies associated with devolution of decision-making authority to local scales. Indeed, the main source of the complexity inherent in decentralization is one of scale. To what extent, and for what services, should central authorities transfer responsibility and resources to local levels in order to most effectively serve the nation?

In this sense, decentralization shares many of the core issues surrounding federalism, which is simply the formal articulation of decentralized governance within a nation's constitution. The diverse configurations of shared sovereignty between central and regional entities that can fall within a federal system of government underscore the complex set of issues that arise in determining the optimal distribution of responsibility and authority between various levels of government (Ferejohn and Weingast, 1997).

The notion of federalism, like that of decentralization, is based on the principle of subsidiarity, which holds that a central authority should play a subsidiary role, performing only those tasks that cannot be performed effectively at a more immediate or local level. This idea dates back at least to the 19th century papal encyclical *Rerum Novarum*, in which Pope Leo XIII sought to define a feasible political economy that would celebrate the autonomy and intrinsic dignity of individual human beings in a way that totalitarian governments did not, yet would recognize the need for public goods and services to support and serve individuals living in community in a way unfettered capitalism never would.

The principle of subsidiarity thus emphasizes the importance of meso-level institutions such as social networks, churches, voluntary associations and local governments. Indeed, it is no accident that enthusiasm for decentralization among development practitioners has paralleled the explosion of interest in 'social capital' within development studies, which holds that social embeddedness plays a central role in the pace and quality of development enjoyed by individuals and communities. Yet much of the social capital literature has projected a relatively naive view of associational life as an unambiguous aid in the process of development, in spite of considerable empirical evidence to the contrary (Platteau, 2000; Barrett, 2005). As we elaborate below, a more nuanced understanding of the process and (sometimes unfulfilled) promise to decentralization requires a more textured and conditional view of the social economics of development, or the ways in which identities and social relations at once constrain and foster different behaviours. Some settings are ripe for decentralization-led improvements in resource allocation and innovation; in others, decentralization can be a siren's call that leads merely to redistribution, often from the relatively powerless to favoured subgroups.

Most observers and analysts would readily concede that the principle of subsidiarity is scale-sensitive. The economies of scale regarding the production of military and defence services, for example, favour national provision by central governments. Central governments are also better placed to correct for inequalities in resource endowments and capabilities across regions. Where localized information and preferences are key determinants of the effectiveness of service delivery (such as in some education or extension services), local governments are likely to be more effective. Governments are nevertheless tasked with the provision of myriad services, and the identification of the optimal scale of decentralization at which to most effectively offer many of these is highly complex.

Despite the sound theoretical arguments in favour of decentralized decision making over public goods and services, the chequered experience of African nations that have recently experimented with decentralization begs for caution and demands a more nuanced, measured and critical prescription of decentralization (Crook, 2003). Several studies examining the evidence of decentralization outcomes among countries in sub-Saharan Africa have failed to document any significant effect of decentralization on poverty reduction or improved local development performance (Adamolekun, 1999; Crook, 2003). While different constraints and dynamics are cited as the factors limiting or negating the potential benefits of decentralization, most explanations point to an underdeveloped socio-economic environment at the micro level, and a lack of supporting institutional infrastructure at the meso level.

For a programme of decentralization to succeed, a set of prerequisites must be met in order for local level institutions to become capable and honest service providers to replace or improve upon central government. Unless local populations are politically mature, have access to adequate information, are aware of their rights and the channels by which they can exercise them, devolving authority to local leaders may result in the capture of resources by an elite few, and undermine the popular support for decentralization. Poorly articulated roles and underdeveloped community- and meso-level institutions necessary to engage the support and resources of key actors such as traditional local authorities, CBOs and NGOs as well as private sector partners can further undermine the incentives and capabilities necessary for local officials and organizations to perform effectively.

The evidence of the outcomes of decentralization in African countries thus far suggests that while decentralization has failed to live up to expectations, and is certainly no panacea for rural empowerment and pro-poor growth, its potential has yet to be fully tapped. For its promise to be met, decentralization has to unfold systematically and sequentially, be attentive to scale-sensitive subsidiarity, have local capacity to efficiently and equitably identify needs and opportunities and allocate resources, build clearly defined meso-level institutions that can effectively link local and central governments as well as other key development partners, and perform necessary facilitative and regulatory functions.

What, then, are the characteristics of the micro- and meso-level socio-economic environment within which a programme of decentralized governance commonly proves welfare-improving? What are the various meso-level institutional constraints that limit or negate the benefits of devolution of authority from the centre? How does the nature of the service to be provided, the development objective sought or the capacity of key actors guide the identification of the optimal scale of decentralization? How do preconditions of successful decentralization differ by context? These are the questions that preoccupy the contributors to this volume. The remainder of this introductory essay sketches out some common threads that run through the subsequent nine chapters.

## 2 Evidence from Kenya: An Overview of the Chapters

Together, the set of empirical studies in this volume illuminates different aspects of decentralization in an attempt to offer a broad understanding of the key issues as well as the challenges and opportunities that must be considered for the design of a coherent, inclusive and ultimately effective programme of decentralization. The volume focuses on the experience of decentralization in rural Kenya and is presented in two parts, which highlight two key, overarching processes that are fundamental in determining the success of decentralization. The context-sensitivity of decentralization outcomes demands detailed attention to the specific socio-economic and geopolitical environment upon which decentralization unfolds and justifies the singular focus on one nation. Nevertheless, the key issues highlighted by the case studies presented herein are based on configurations of objectives, capabilities and constraints replicated across numerous rural environments across Africa, indeed throughout much of the rest of the low-income world. As such, lessons generated from these studies can be readily generalized and applied to inform decentralization efforts in similar contexts across the African continent and beyond.

### 2.1 Successes and failures of decentralization in rural Kenya

Part I, comprising Chapters 2–6, examines the successes and failures of select decentralization efforts across Kenya's rural landscape. It investigates the effects of decentralization in settings spanning devolution in the administration and management of cooperatives, the provision and financing of agricultural extension services, and the authorizing of local governments and community leaders to adjudicate over natural resources management and associated land use conflicts. It offers evidence of the various forms that decentralization can take and the varying outcomes it can yield. By studying the institutional and socio-economic context by which decentralization took place in each case, the chapters offer insights into the challenges and correlates of successful decentralization.

In Chapter 2, Nyoro and Ngugi study the factors associated with the success and failure of agricultural cooperatives in the liberalized Kenyan economy in which government withdrew the administrative and financial support it had previously offered cooperatives and devolved governance authority to cooperative members. Giving producers greater or complete ownership and autonomy over their cooperatives was expected to increase both their productive and marketing efficiency. As Nyoro and Ngugi document, the retreat of government resulted in a legal and regulatory vacuum that adversely affected the performance of many cooperatives and facilitated exploitative behaviour on the part of some unscrupulous members and business associates. Furthermore, a lack of general understanding among the membership of the business of running a cooperative left some with a dearth of management, accounting and marketing skills necessary to compete effectively. The experience of decentralization of control over cooperatives was not uniformly negative, however, as some evolved in response to the changing environment and seemed to thrive, delivering good value to their members. Nyoro and Ngugi's results show that successful cooperatives were generally characterized by a well-educated and skilled management committee, and transparency, accountability and responsiveness of the management to the members at large.

In Chapter 3, Mude continues the exploration of the effects of decentralization within agricultural producer cooperatives. He identifies deficiencies in the formal and informal institutional environment underlying coffee cooperatives that undermined the efficacy of market decentralization in the smallholder coffee sub-sector in Kenya. Mude describes how the free and unregulated access to cooperative coffers enjoyed by elected cooperative officials under the decentralized regime increased the incentives for corrupt members to manipulate the electoral process in a bid to win cooperative elections. Along with a politically immature and easily misinformed membership, the traditional voting practice of lining up behind one's favoured candidate facilitated rampant vote-buying by offering corrupt candidates a costless mechanism to enforce illicit vote-buying arrangements. Having won elections, cooperative officials could rely on the law that required members to market their coffee exclusively through their designated cooperative to extract maximum rents from their members. Furthermore, despite widespread but unsupported claims that economies of scale favour large cooperatives, Mude's analysis shows that small cooperatives outperform larger ones, on average. Scale effects exist at the factory level at which physical processing takes place, but not at the larger cooperative level (there are multiple factories in most cooperatives) where size facilitates corruption more than coordination. Such countervailing scale effects reveal structural deficiencies in cooperative organizations that could hijack devolution reforms.

In Chapter 4, Mugunieri and Omiti explore whether decentralization of extension services enhances access for poor farmers, especially for demand-driven extension services that respond directly to farmers' needs, and encourages active farmer participation. They document how decentralization has been manifest through sharing agricultural extension provision responsibilities among a coalition of local government, CBOs, NGOs and private sector partners, with the possibility of user fees to generate pay-as-you-go revenue streams at local level. Using primary data collected in eastern Kenya, they again show that meso-level institutional environment is a critical determinant of successful decentralization. They find that farmer access to, and use of, extension services increases with density, presence and involvement of CBOs and NGOs in services delivery, although there remains a crucial role for government – whose extension services are widely regarded as of highest quality – in strategic areas of a more 'public good' nature.

In Chapter 5, Nambiro and Omiti examine client assessment of quality, cost and accessibility, as well as willingness to pay for extension services in rural western Kenya. Their novel willingness-to-pay estimates for extension services are especially important in that they indicate a real possibility for some decentralized resource mobilization to complement and facilitate decentralized resource allocation and more demand-driven extension services provision. A majority of extension clients continue to rely heavily on government-provided extension services and indicate that they prefer those services, on both quality and affordability grounds, over alternative suppliers. This is not surprising considering pervasive levels of rural poverty that make it difficult to purchase non-essential services such as extension. None the less, half of the households express a willingness to pay for extension services, at a rate that would seem to offer promise for improved local financing of extension services provision.

In Chapter 6, Munyao and Barrett show that creating local government institutions as a branch of the national civil service and investing such institutions and representatives with formal authority traditionally exercised by indigenous institutions can inadvertently marginalize local populations. Munyao and Barrett describe the experience of Gabra transhumant pastoralists in northern Kenya, who migrate seasonally with their livestock in search of water and fodder. Increasing sedentarization and the influx of migrant settlers into the Hurri Hills, the Gabra's traditional wet-season range lands, resulted in increasing land conflicts between settlers and the Gabra, while the creation of local government with formal authority over land rights subtly shifted power from Gabra pastoralists to Gabra settlers and other immigrants. Local government disproportionately favoured settlers by virtue of their permanent presence in the Hurri Hills. The authors caution that without a keen understanding of the array of local interests and the prospect for competing centres of power within local jurisdictions, and without buttressing effectively functioning traditional institutions, devolution of formal power may lead to an inequitable and unsustainable distribution of authority and resources that may serve to fuel, rather than resolve, land and resource conflicts.

## 2.2 Socio-economic and institutional preconditions for successful decentralization

Highlighting the importance of a conducive and receptive socio-economic environment at the local level as a precondition for successful decentralization, Part II focuses on the social networks, informal groups and community organizations that can act as a vehicle by which administrative authority is effectively devolved to local-level institutions and through which the potential for abuse can be either checked or fostered. Because so much of the outcome of decentralization experiences appears to turn on the pre-existing condition of meso-level informal institutions, what Barrett (2005) terms the 'social economics of development' becomes a crucial determinant of performance. Decentralization cannot be introduced into an information or managerial capacity vacuum. Communities must have the wherewithal to impose standards and demand accountability and performance from local leaders. Communities must also have internal mechanisms to effectively resolve intracommunity conflicts and disagreements. A key to effective decentralization to empower the poor is increased, broad-based participation in local decision making concerning common pool resources and public goods and services provision.

In Chapter 7, Haro, Doyo and McPeak provide an excellent and promising example of how inclusive and flexible community-led initiatives can effectively deal with pressing problems and lead to the establishment of a trusted, effective conflict and resource management organization. They describe the iterative process by which a community initiative tasked with natural resource management (NRM) in northern Kenya adapted its mandate to first tackle the source of the problem – conflicts stemming from overlapping claims on land and resource use – before it could address NRM issues effectively. Despite the varied set of interests and the heterogeneity of the affected communities, conflicts were resolved and a programme of sustainable and equitable natural resource use was later adopted. So effective did the organization become in identifying and solving local problems that government sought its assistance in resolving broader issues of insecurity and violence in the region.

In stark contrast to the example of a successful community-led organization presented by Haro, Doyo and McPeak, Osterloh and Barrett, in Chapter 8, examine the unfulfilled promise of financial service associations (FSAs), a particular form of microfinance institution, and highlight the unintended, detrimental effects that they have had on their poorer members in Kenya. A form of microfinance institution in which equity capital for the loans fund is generated by selling shares to members, FSAs are designed to be locally sustainable but are often externally introduced. Osterloh and Barrett expose the design of FSAs to be based on untenable assumptions of the social motivations and incentives of the locally chosen board elected to manage the FSA. They argue that FSAs fail because the informational advantage of local monitoring and enforcement is trumped by social incentives to peddle favours in the form of approving loans expected not to be creditworthy. This situation persists because members at large are unaware of the loan review process and of the effect that the unsustainable rate of default on loans has on the value of their shares, because individual credit committee members personally bear the brunt of disgruntled prospective borrowers, as FSAs routinely fail to follow established accounting procedures. The result is that FSAs systematically disadvantage poorer members of communities and effectively redistribute resources from poorer FSA members to richer ones who more commonly default at larger scales, eroding the value of wealth held in the form of FSA shares.

In Chapter 9, Amudavi explores the determinants of group participation and the differential impact of group membership on members' welfare in three districts in central and western Kenya. He finds that even when controlling for other key covariates, group participation statistically significantly improves welfare. A related key finding reveals that group characteristics matter and, specifically, only those groups that are supported by, or have links to, external organizations (what Amudavi calls 'supra groups') confer demonstrable welfare benefits to their members, while internally oriented groups (what Amudavi calls 'community groups') fail to deliver similar gains. Supra groups are better endowed, offer a range of extension support and training to their members, and provide access to financial services, markets and improved livelihood opportunities. Amudavi also finds that well-endowed households are more likely to gain membership in supra groups, a phenomenon that may lead to rising intra-community inequality. Thus, policies promoting group formation should pay close attention both to the capacity of the group to establish valuable external linkages for members and to explicit mechanisms to ensure the involvement of a community's poorer members.

In Chapter 10, Hogset explores mechanisms of bilateral financial transfers within social networks in central and western Kenya. She finds that participation in transfer networks is largely dependent on one's resources. Wealthier individuals both give and receive larger transfers, and these transfers are more likely to be in cash rather than in kind. She also finds that women are more likely to engage in transfers, especially those in kind, and that transfers are most often given for consumption purposes rather than for investment, except for children's educational expenses that attract the largest transfers. No evidence is found that social networks support households with seriously ill members, except for funeral expenses, suggesting considerable uninsured exposure to health shocks, even for those with reasonably extensive social networks. These insights into the functioning of voluntary, decentralized social networks illuminate prospective issues that can arise with decentralization of more formal services provision. For example, the finding that recent immigrants and the poor rarely have access to meaningful networks suggests that unless decentralized programmes explicitly emphasize service to these sub-populations, they are likely to be inadvertently excluded from any benefits resulting from decentralization – as they are from voluntary social networks – and thereby further marginalized.

## 3 Cross-cutting Lessons for Future Decentralization Initiatives

Together, the nine chapters that make up this volume offer some key lessons for policy makers considering a programme of decentralization and for scholars undertaking further research on the determinants of successful decentralization initiatives. In this section, we briefly highlight the main insights offered by the empirical studies that follow.

Elite capture, whereby well-connected individuals secure positions of advantage in decentralized organizations in order to direct benefits towards themselves or even to misappropriate resources, has been widely identified as a key risk of decentralization (Bardhan, 1997). Measures limiting the prospective emergence and impact of elite capture are thus a critical component of efforts to effectively devolve power and resource management. The first key lesson we emphasize involves such measures. Several chapters in this volume show that while transparency of decision making by those wielding authority is critical, the anonymity of key decisions is likewise important to protect decision makers from manipulation. Mude (Chapter 3) argues that election capture among coffee cooperative board members is facilitated by a system of open voting which supports vote-buying by fully disclosing members' votes. Once corrupt individuals capture the board, a lack of accountability and transparency in board decisions makes it relatively easy to expropriate the organization's resources and exploit members. Non-anonymity of decision making can also negatively affect microfinance performance. As Osterloh and Barrett (Chapter 8) argue, members elected to make crucial lending decisions for FSAs rarely refuse loans to applicants or punish those who delay or default in repayments. Again, elite capture is exacerbated by a lack of transparency and accountability of FSA affairs, in part made possible by too-open decision making, which creates socially undesirable incentives for individual actors.

These examples suggest that while transparency of process and structures for accountability are critical, openness is not everywhere and always desirable, particularly not where openness provides information that can be used by negatively affected parties to punish decision makers for public-minded actions they take in the interest of the organization as a whole. Anonymity of such decisions coupled with transparency in operations and decision-making rules seems likely to reduce the incidence of elite capture by making it more difficult to siphon group resources for personal use.

A second major obstacle limiting the success of decentralization is the dearth of well-trained, knowledgeable, open-minded managers capable of running successful institutions in rural Kenya. Nyoro and Ngugi (Chapter 2) emphasize the importance of highly educated managers with integrity. Osterloh and Barrett (Chapter 8) similarly argue that FSA managers may be less inclined to overlook their contract enforcement responsibilities if they were well trained in financial accounting. The type of managerial acumen necessary for success is dependent on the objectives and mandate of the organization in question. Haro, Doyo and McPeak (Chapter 7) offer a great example of how the engineers of the effort to resolve resource management conflicts among communities with overlapping ownership claims succeeded despite the sensitivity of the issue they were trying to address. A genuine commitment to uncover the main source of the problem without attributing blame, an ability to adapt or change course in the face of new information, and the inclusion of all affected parties in discussion forums defined the managerial skills employed to yield a positive resolution.

An equally important and complementary prerequisite for success in decentralization is the political maturity and active involvement of the membership. If the members are ill-informed and passive, they can be easily manipulated and divided in their demand for accountability and effective service from their leaders. One way to spread awareness among rural communities and empower residents to proactively engage their leaders and representatives is to encourage competition, diversity and partnership in service provision and market involvement. As Mude (Chapter 3) argues, the law constraining smallholder coffee producers to market their produce only through the cooperative to which they belong allows board members to drive the prices they offer their members below what would otherwise prevail in a competitive market. The legally enforced local monopsony discourages members from seeking better opportunities as any threat to market elsewhere would not be credible. Nyoro and Ngugi (Chapter 2) explain that allowing individual middlemen to buy milk directly from producers accelerated the collapse of ineffective dairy cooperatives while forcing survivors to improve the terms they offered their clients. Mugunieri and Omiti (Chapter 4) also present evidence that competition among service providers, in their case in agricultural extension, likewise improves quality of, and access to, services.

The analogue for competition in self-help, networking or service provision groups is engaging in partnerships that reach beyond the confines of villages to effectively couple the coordinating capacity of homophilous, informal, local groups and networks with the resources accessible through external agencies. Amudavi (Chapter 9) finds statistical evidence of the benefits of external linkages in group performance. He shows that welfare increases statistically significantly only for individuals who are members of groups that are supported by external organizations. Haro, Doyo and Mcpeak (Chapter 7) present an example of a grass-roots movement that owes its success to the active inclusion of various affected communities, government officials, local organizations and external NGOs. Such partnerships, drawing from the expertise and objective of different entities, enabled all affected communities to iron out their differences and reach a mutually beneficial solution.

While the involvement of external agencies with CBOs or public–private partnerships in service provision often incubates an effective and equitable programme of decentralization, this can only be true if all affected sub-populations are adequately included in the process. Munyao and Barrett (Chapter 6) show how Gabra pastoralists lost access to their traditional grazing lands in favour of settlers and immigrants who had better access to local government officials recently empowered through a process of deconcentration. This also undermined traditional Gabra authorities. Hogset (Chapter 10) shows how poorly endowed households are routinely excluded from social networks that offer opportunities for accessing informal finance and insurance from associates. At a more formalized level, Amudavi (Chapter 9) finds that wealthier households are more likely to be members of beneficial community organizations. Results of willingness to pay for extension services in Nambiro and Omiti (Chapter 5) similarly caution that private, profit-based, extension delivery may limit the extension services available to poor farmers, who are systematically less able or willing to pay for services for which wealthier neighbours are demonstrably willing to pay. Mugunieri and Omiti (Chapter 4) highlight various novel configurations of extension provision, including farmer field schools, community field days and certain private–public partnerships that could insure that the increasing trend towards extension commercialization does not discriminate against poor farmers.

These key lessons, drawn from specific case studies and analyses looking at different experiences of decentralization in rural Kenya, can usefully inform the design of inclusive, pro-poor programmes of decentralization in Kenya and elsewhere. Decentralization is a multidimensional and complex process. Effective decentralization to accelerate broad-based growth and welfare improvement in rural areas is clearly possible but by no means automatic.

## References


Adamolekun, L. (ed.) (1999) *Public Administration in Africa: Main Issues and Selected Country Studies*. Westview Press, Boulder, Colorado.

Bardhan, P. (1997) Corruption and development: a review of issues. *Journal of Economic Literature* 35(3), 1320–1346.

Barrett, C.B. (ed.) (2005) *The Social Economics of Poverty: Identities, Groups, Communities and Networks*. Routledge, London.

Crook, R.C. (2003) Decentralisation and poverty reduction in Africa: the politics of local-central relations. *Public Administration and Development* 23, 77–88.

Ferejohn, J. and Weingast, B.R. (eds) (1997) *The New Federalism: Can the States Be Trusted?* Hoover Institution Press, Stanford, California.

Platteau, J.P. (2000) *Institutions, Social Norms, and Economic Development*. Harwood Academic Publishers, Amsterdam, The Netherlands.

Prud'homme, R. (2003) Fiscal decentralization in Africa: a framework for considering reform. *Public Administration and Development* 23, 17–27.

Smoke, P. (2003) Decentralization in Africa: goals, dimensions, myths and challenges. *Public Administration and Development* 23, 7–16.

World Bank (1999) *Entering the 21st Century: 1999–2000 World Development Report*. The World Bank: Washington, DC.

# 2 A Qualitative Analysis of Success and Failure Factors of Agricultural Cooperatives in Central Kenya

J.K. Nyoro and I.K. Ngugi

## 1 Introduction

The performance of smallholder producer organizations impacts on the economic and social welfare of the poor in countries such as Kenya. Cooperatives, defined as user-owned, user-controlled and user-benefiting organizations whose objective is to promote the economic interests and general welfare of members in accordance with cooperative values and principles (Cook, 1997; USDA, 1997), are the most common form of producer organizations in Kenya. Cooperatives may be agricultural, non-agricultural, unions or Savings and Credit Cooperatives (SACCOs). This study focuses on agricultural cooperatives in Kenya.

In Kenya, cooperatives play a major role in organizing the production, processing and marketing of crop and livestock commodities. Cooperatives also play a critical role in the provision of savings and credit services, especially for the rural poor who are largely unable to access formal financial services. By pooling the meagre resources of small-scale producers, cooperatives can harness economies of scale to improve the productivity, and ultimately the welfare of members. The benefits of collective organization that cooperative members might enjoy include improved access to credit and extension services, and greater bargaining power to leverage for more competitive prices and advocate for more advantageous policies. By 1999, Kenya had over 9000 registered cooperatives out of which 46% were agricultural and 38% were savings and credit, supporting over 2.5 million of Kenya's nearly 35 million people (Republic of Kenya, 2002). Among Kenya's key commodity exports, 100% of cotton, 87% of pyrethrum and 52% of coffee are sourced from cooperatives (ICA, 1995).

With poverty most prevalent in rural areas and the rural economy largely based on agricultural production, economic growth and poverty alleviation are inextricably linked to agricultural sector performance, especially in smallholder agriculture. Indeed, according to the nation's Poverty Reduction Strategy Paper

 *Decentralization and the Social Economics of Development* (eds C.B. Barrett, A.G. Mude and J.M. Omiti)

(PRSP), agriculture and rural development is identified as the number one priority for poverty alleviation and economic growth (Republic of Kenya, 2002). Given that the bulk of commercialized smallholder production is associated with cooperatives, improving their performance is crucial to the alleviation of poverty and stimulation of economic growth.

Since the 1980s, liberalization, globalization and deregulation have created an enormously challenging environment for Kenya's agricultural cooperatives. The structural adjustment era of the 1990s was characterized by major reforms to liberalize various segments of the economy and this greatly impacted their performance. Structural adjustments included internal changes (notably privatization and deregulation) as well as external ones, especially the reduction of trade barriers. Responding to liberalization pressures for minimal government involvement in the market process, the government drastically cut its support of cooperatives. The government ceased to support the cooperative movement through direct assistance and subsidized services in 1998 when a new Cooperative Societies Act and Sessional Paper No. 6 of 1997 on cooperatives in a liberalized economy became effective (Republic of Kenya, 2002). For dairy cooperatives, this meant that government no longer subsidized the provision of veterinary and artificial insemination services, which are costly and yet critical for building a large stock of healthy, high-output dairy cattle. Anecdotal evidence suggests that the use of both services has plummeted, leading to a decline in the fraction of top-quality pure-bred dairy cattle, and hence a fall in average lactation volumes.

The steady decline of government support for cooperatives culminated in the new Cooperatives Act of 1998 wherein the government relinquished all policy-making jurisdictions over the economic activities of cooperatives and took on a minimal advisory and regulatory role. Where government previously provided extension support as well as accounting and auditing services, cooperatives are now required to source such services on their own. The lack of regulatory oversight has paved the way for seemingly increased rent seeking, particularly among coffee cooperatives whose substantial revenues make corruption a lucrative strategy for unscrupulous self-interested board members (Mude, Chapter 3, this volume). As leadership positions became highly coveted, political feuds, many of them characterized by violence and sabotage, became distressingly commonplace. Without government oversight, mismanagement and neglect often took root, and critical infrastructure, such as roads and pulping stations, were too often left to deteriorate. Although coffee cooperative members pay a road maintenance levy, most roads under the jurisdiction of cooperatives continue to be in poor condition. During wet seasons, these roads become impassable, thereby leading to long delays in the delivery of harvested produce. The quality of coffee, milk and other perishable commodities rapidly deteriorates as a function of the time lag between harvest and processing.

Liberalization also brought with it the legalization of intermediary traders in the dairy industry and other sub-sectors. Individuals could obtain licences to purchase milk at the farm gate and sell it fresh to consumers or to processors for pasteurization. Unlike coffee farmers legally obliged to sell their output to a cooperative, members of dairy cooperatives can legally sell milk to hawkers while maintaining active membership with their cooperative. This sudden inflow of competitors to the market crippled many dairy cooperatives that were unable to offer members competitive contracts. Lower fixed costs and a more flexible operating structure allowed hawkers to offer more favourable terms, such as prompt payment and better prices.

Meanwhile, historically low international coffee prices have contributed to a stagnant Kenyan coffee sector. Low international coffee prices are caused by imbalances between supply and demand. Between 1990 and 2002, global coffee production increased at an annual rate of 3.6% while demand grew by merely 1.5%, leading to price collapse in the face of price inelastic demand (ICO, 2003). The effect of low international prices trickles down to farmers in the form of lower payments. By increasing the relative opportunity cost of coffee production, and by reducing the revenue available for purchasing quality inputs in the subsequent season, low prices contribute to reduced output in the ensuing season.

Many of the problems that beset smallholder agricultural cooperatives arise and persist as a consequence of a weak legal and regulatory framework that rarely and inconsistently enforces contracts and punishes those who breach contract. This allows corrupt and manipulative behaviour to blossom and creates an environment of distrust not conducive to productive economic activity. Widespread anecdotal evidence points to the harm done by the rent-seeking behaviour made possible by a weak regulatory structure and rarely enforced rules. In principle, well-managed cooperatives can resolve, or at least attenuate, these problems.

However, confident in the knowledge that members have little legal recourse for justice, corruption among the management and staff of cooperatives too often occurs. This can take any of several forms. Overstating deliveries or recording of fictitious deliveries is common as is collusion with suppliers to manipulate procurement figures. Loans have been known to be offered to applicants who did not satisfy minimum requirements and outstanding loans waived as favours to friends or allies. The outright embezzlement of funds is not uncommon. A weak regulatory environment also makes cooperatives vulnerable to exploitation by deceitful businessmen.

While these challenges are formidable, failure is not foreordained. On this difficult new playing field, a few cooperatives have been doing well even while others have been liquidated or are on the verge of collapse. The factors behind the success or failure of such cooperatives have, however, not been thoroughly investigated. This chapter explores the factors that distinguish between cooperatives that have proved successful and those that have failed in the liberalized, globalized and deregulated environment of Kenyan agriculture in the last decade or so. In particular, we identify differences in organizational structure and other attributes that result in the diverging performances of cooperatives in the coffee and dairy sub-sectors in a high potential area of central Kenya.

# 2 Methods

## 2.1 Study area

The study was undertaken in the Kiambu district of Kenya's Central Province. With an absolute headcount poverty level of 25%, Kiambu has the lowest poverty rate among all districts in Central Province (IEA, 2002). With a total population of 744,000 and density of 560 persons per square kilometre, it is also the most populous and densely populated district. In Central Province, Kiambu has the fifth highest monthly mean household income (US$56.7)[1] and fifth lowest unemployment rate (8.8%). The enrollment rate for primary school level in Kiambu is 72.6% while that of secondary level is 43.5% (IEA, 2002).

Kiambu is home to a large number of cooperatives. Table 2.1 shows the different types, number, membership, share capital and financial turnover of cooperatives in the district. The active cooperatives had normal business going on throughout the year. Delays in payments discouraged farmers from full participation, thereby causing the number of active cooperatives to go down in 2003 (DCO, 2003). Such was the case in the pyrethrum sector, where the Pyrethrum Board of Kenya delayed payments to cooperatives, which in turn delayed payments to members. SACCOs also were characterized by reduction in the number of active cooperatives. Most of those dormant were small SACCOs, possibly because they could not benefit from economies of scale. Further, non-remittance of deductions by employers to the SACCOs was prevalent (DCO, 2003). Table 2.1 shows that dairy followed by coffee were the dominant cooperatives in the district in terms of membership. The share capital[2] was, however, higher with SACCOs than with agricultural cooperatives though the financial turnover[3] was highest with dairy cooperatives.

In addition to these categories of organizations, others that are active in the district include one cooperative union (Kiambu Coffee Growers Cooperative Union), one vet services cooperative, one land purchasing cooperative, one women's cooperative and 14 multipurpose cooperatives that not only deal in commodity marketing but also in other activities such as land purchase (DCO, 2003).

Although there are many types of cooperatives in the district, this study purposively focuses on the coffee and dairy societies, the leading agricultural cooperatives in the district in terms of numbers, membership, share capital and financial turnover. In the coffee sector, cherry intake in 2003 was 18 million kilograms, which fetched US$2.8 million, of which 59% was paid to members at an average rate of KSh 8 (US$0.11) per kilogram of cherry. In the dairy

[1] Exchange rate: Kenyan shilling per US dollar (KSh/US$) = 75.9 (2003), 78.7 (2002) (World Fact Book, 2004).

[2] Share capital in this context refers to the amount of money paid by members when joining cooperatives. The figures used in this study were estimates by the District Cooperative Office as reported in their annual reports.

[3] Turnover refers to the amount of money that flows through the cooperatives per year as estimated by the District Cooperative Office.

**Table 2.1.** Cooperatives in Kiambu district, 2003. (From Ministry of Cooperative Development and Marketing, Kiambu District, 2003.)

| Cooperative type | Total cooperatives | Active cooperatives | Number of members | Share capital (US$) | Turnover (US$) |
|---|---|---|---|---|---|
| Coffee societies | 9 | 8 | 26,096 | 574,898 | 1,247,667 |
| Dairy societies | 15 | 14 | 41,732 | 501,785 | 15,465,612 |
| Pyrethrum societies | 6 | 3 | 10,692 | 17,196 | 16,208 |
| Urban SACCOs | 138 | 77 | 12,299 | 7,844,533 | 1,173,465 |
| Rural SACCOs | 24 | 14 | 16,501 | 9,074,422 | 1,867,559 |
| Housing societies | 7 | 3 | 4,150 | 1,135,504 | 47,809 |
| Union (KDPCU) | 1 | 1 | 21 | 1,417,407 | 206,236 |

sector, milk intake during 2003 was 50 million litres, which fetched US$11.6 million, of which 86% was paid to members at an average rate of KSh 14.60 (US$0.19) per litre of milk.

## 2.2 Choice of area, cooperatives and measures of success

Key informants were involved in selection of the study area and appropriate cooperatives for investigation. The informants were selected from the Coffee Board of Kenya management team, the Ministry of Cooperative Development and Marketing, and the Kenya Union of Savings and Credit Cooperatives Nairobi Regional Office. Kiambu was identified by key informants as having both the best and worst performing cooperatives.

Key informants suggested two successful and two unsuccessful cooperatives from each sub-sector. The choice of successful cooperatives was common among the key informants but they gave varied suggestions on the unsuccessful ones. A simple random sampling technique was therefore used to select two among the pool of suggested unsuccessful cooperatives. The selected ones were then studied in more detail to identify characteristics of success and failure. Although there exists a key difference between measures of success (e.g. profitability) and causal factors (e.g. strong leadership), in this study we assume the difference is blurred and literally adopt the key informants' choice of the indicators. The measures were determined *ex ante* while the causal or associated factors of success or failure were researched.

## 2.3 Data collection

Primary data were collected in structured interviews of the key informants and members of the management committee of the selected cooperatives. A largely open-ended questionnaire was fielded to elicit qualitative responses. A separate questionnaire to collect quantitative data was presented to the cooperative

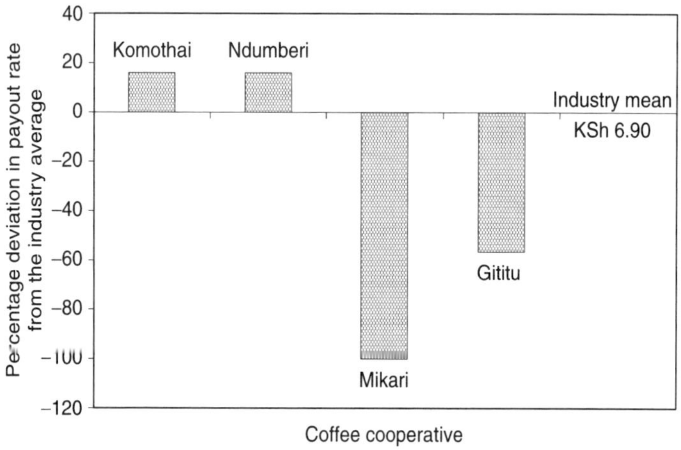


**Fig. 2.3.** Coffee cooperatives' payout levels in relation to industry rate (2003).

However, the average 2003 payout levels are shown in Fig. 2.3. Confirming the key informants' categorization, the coffee cooperatives classified as successful – Komothai and Ndumberi – paid more than the industry average (KSh 6.90/kg) while the unsuccessful cooperatives paid below the local industry mean.

In terms of promptness of payments, as in dairy, the successful coffee cooperatives paid farmers more promptly. On average, Komothai and Ndumberi paid farmers after 3 and 6 months, respectively. On the other hand, the poor-performing Mikari and Gititu each paid after 9 months. In addition to paying promptly and at high rates, Komothai and Ndumberi had SACCOs and stores that enabled them to advance credit to farmers in form of finance as well as farm inputs and were not characterized by wrangles and hostilities, unlike the unsuccessful coffee cooperatives (see details below). The performance of the cooperatives on the basis of the four indicators therefore confirms the classification of Komothai and Ndumberi as relatively successful, and Mikari and Gititu as relatively unsuccessful coffee cooperatives.

## 3.2 Attributes of successful cooperatives

What are the factors that account for the success of cooperatives in Kiambu? The results of the qualitative analysis suggest that the factors associated with successful cooperatives are vertical integration, high-quality produce, appropriate skills and education of management committee and staff members, timely and appropriate information, diversification into profitable enterprises, large number of members, large quantity of produce marketed through cooperatives, proper record keeping and level of technology, SACCOs and stores for farm inputs.

### *3.2.1 Vertical integration*

Most successful cooperatives in the selected sample were characterized by some degree of vertical integration. Good examples in this study were Limuru and Githunguri in the dairy sub-sector and Komothai in the coffee sub-sector. The benefits to vertical integration include, but are not limited to, savings on transport and processing cost, gains from value addition, and gains from the sales of by-products such as coffee husks.

Komothai is a good example of a cooperative with benefits to vertical integration in the coffee sector. Before setting up the milling plant in 2001, Komothai would transport its pulped coffee to Kenya Planters Cooperative Union (KPCU) for hulling and marketing. After the construction of its milling plant, only the clean coffee (without husks) was transported to KPCU for marketing, resulting in substantial savings to Komothai. The cooperative was able to charge its members US$50 per tonne of parchment (without pulp) for hulling, handling, quality analysis and colour sorting, while KPCU charged US$85 for the same services (70% higher). In addition, by milling its own coffee, the management indicated that milling losses decreased by at least 30% compared to use of a commercial miller. Furthermore, when Komothai used to deliver its coffee to KPCU for milling, the channel was characterized by a long wait due to the large number of cooperatives that KPCU served. This translated into delays in paying farmers. Once Komothai constructed their own milling plant they were no longer constrained by the delays at KPCU and were able to reduce the average time between delivery of parchment and payment to farmers from 8 to 3 months.

Milling their own coffee also allowed Komothai to generate some income by selling off coffee husks, a by-product of the milling process. Sales of coffee husks also benefited the farmers who used them as a cheap substitute for cooking fuel and as material for making farmyard manure. Moreover, free of husks, the weight of delivered coffee, conditional on value, is much lower, resulting in decreased transportation costs.

Vertical integration also played a role in keeping Komothai intact during the liberalization era. Liberalization broke the existing monopoly of KPCU in milling of coffee. New millers emerged, including Thika Mills, Socfinaf and Gatatha. This caused many large cooperatives with multiple factories (pulping stations) to split due to irreconcilable differences in the choice of preferred miller by farmers affiliated with different factories. As Komothai had its own milling plant, they did not face the same problem and remained intact. This saved Komothai from the significant financial and political costs involved with splitting up a cooperative. This stability lured other farmers to join the cooperative, thereby increasing production and sales, reducing overhead costs per kilogram processed and improving the potential profitability of Komothai cooperative.

In the dairy sub-sector, vertical integration was noted with Limuru and Githunguri, both of which processed and packaged milk and other dairy products in addition to collection and marketing. Githunguri processed and packaged fresh milk, yogurt, ghee and butter. They had plans to begin processing *mala* (fermented milk) within the next year and had already installed the necessary facilities when we interviewed them. In addition to these products, Limuru were already processing *mala*. The value added from processing and packaging their own dairy products contributed to the higher payments that Limuru and Githunguri were able to pay their members. In addition to higher payout rates, the delays in payment that were common when deliveries were done through commercial processors (that handled milk for many cooperatives) were no longer observed.

### *3.2.2 High-quality produce*

All else being equal, high-quality produce yields higher prices. Many factors influence the quality of coffee and dairy products. For coffee, farm-level factors are one of the main determinants of quality. These include soil fertility and acidity as well as the quality and timing of fertilizer and pesticide application, weeding, pruning, harvesting, the extent of intercropping and other such husbandry practices. At the pulping and milling stages, proper storage and transportation, a hygienic fermentation process, and efficient sorting of beans is vital to maximizing the quality of the harvest. According to the management, a *debe*[5] of coffee berry from well-tended coffee plants weighs 13–18 kg and is of better quality, while that from a poorly managed farm weighs only about 7 kg and is of poor grade. In the dairy sector, the quality of milk is influenced mainly by the care or hygiene observed in handling the product. As milk is highly perishable, quick and reliable methods for transporting fresh milk to processing plants are vital. Reliable storage and regular quality testing is necessary to limit spoilage and guard against the sale of milk adulterated by adding water or other impurities, or milk drawn from sick cows.

In both the dairy and coffee sub-sectors, several strategies geared to improving the quality of their output were employed. In our sample, the successful cooperatives were more likely to pursue quality control measures than were the unsuccessful cooperatives. Table 2.2 lists some of the various strategies used by our sample cooperatives to improve quality. These include use of field committees that supervise farm management practices, strict vetting of personnel who directly handle coffee parchment at factory level, use of modern drying bins in coffee factories, use of informal workers at milling plants to do hand-sorting of unwanted beans that would otherwise lower the quality of the whole batch, regular and impromptu checks on milk at collection points and the use of milk-testing instruments at collection centres, cooperatives and processing plants.

### *3.2.3 Qualification of management committee and staff members*

Successful cooperatives had staff and management committees with relatively higher educational attainment than the unsuccessful ones (Table 2.3). The highly qualified personnel were in managerial and bookkeeping positions. Githunguri dairy employed several university graduates and their internal auditor was already pursuing a Master's degree. On the contrary, unsuccessful cooperatives like Nderi had only three employees and none of them had been educated beyond form four. The education and skills of the management committee and staff when used properly are likely to promote success of the cooperative.

[5] A *debe* is a metallic or plastic container commonly used at farm level to measure the quantity (in terms of volume) of coffee harvested by an individual.

**Table 2.2.** Quality control strategies by cooperatives.

| Sub-sector | Performance | Cooperative | Quality control strategies |
|---|---|---|---|
| Coffee | Successful | Komothai | Elected field committee that supervises weeding, pruning and intercropping<br>Casual employees who do effective hand-sorting of unwanted coffee beans at milling plant |
| | | Ndumberi | No intercropping with heavy crops (e.g. maize and napier grass)<br>Use of well-aerated wire mesh bins rather than concrete bins in drying parchment<br>Strict on timely delivery and pulping |
| | Unsuccessful | Mikari | Only recently:<br>Institute inspection committee to oversee farm input credit fungibility<br>Train farmers on benefits of good farm management<br>Use non-smokers and perfume-free casuals in drying parchment to ensure non-contact since smoke/butt and perfume adversely affect quality<br>Engage an experienced and quality-sensitive manager |
| | | Gititu | None |
| Dairy | Successful | Githunguri | Lactometer and testing guns at collection centres<br>Special committees to do impromptu checks<br>Testing instruments at plant |
| | | Limuru | Impromptu checks at collection centres<br>Lactometers |
| | Unsuccessful | Nderi | Organoleptic tests |
| | | Kinale | Organoleptic tests<br>Occasional inspection |

**Table 2.3.** Qualifications of cooperatives' management committee and staff.

| Sub-sector | Performance | Cooperative | Most qualified committee member | Most qualified staff member |
|---|---|---|---|---|
| Dairy | Successful | Githunguri | University | Postgraduate |
| | | Limuru | University | University |
| | Unsuccessful | Nderi | Secondary | Secondary |
| | | Kinale | Secondary | Secondary |
| Coffee | Successful | Komothai | Diploma | University |
| | | Ndumberi | Polytechnic | Polytechnic |
| | Unsuccessful | Mikari | University | Secondary |
| | | Gititu | Secondary | Polytechnic |

### *3.2.4 Timely and appropriate information*

The four successful cooperatives were likely to hold meetings with members more frequently, especially when there were controversial issues to be dealt with. Holding frequent meetings has the benefit of allowing members to participate more actively in the decision-making process of the management. This also forces the management to be more accountable to its members. Frequent meetings allow the dissemination of important information, the exchange of ideas and also provide a forum for the resolution of any disagreements that may arise. In addition to annual general meetings (AGMs), immediately after liberalization, Komothai held several meetings with members at respective factories. In so doing, they were able to convince the members of the benefits of not splitting and of building their own mill, which further sustained the unified cooperative. The meetings were timely in that they were done just before profit-motivated millers got into the area. The new millers had the strategy of targeting cooperatives that had more than one factory and attempting to influence members to split along factory lines and deliver coffee beans to them. While these efforts by millers led to the split in some cooperatives (e.g. Gititu), the timely communication by successful cooperatives like Komothai and Ndumberi was effective and helped the cooperatives remain intact, thereby continuing to benefit from economies of scale and maintaining high payout rates, unlike Gititu, which split and was characterized by wrangles and delayed payments. Githunguri dairy had a habit of holding frequent (zone-based) meetings at collection centres to address issues that were zone-specific, thereby improving communication. Some successful cooperatives (Ndumberi and Limuru) had suggestion boxes that were important in airing views that members were afraid of sharing publicly. Further, to improve information access by farmers, it was common for successful cooperatives to organize demonstration days and other training that informed farmers. Unsuccessful cooperatives largely relied only on AGMs in communicating information to members. Effective communication and training curtailed complaints and subsequently promoted harmony amongst members and cooperatives' management, and hence curbed wrangles in the successful societies.

Timely and effective communication is beneficial not only to farmers but also to management committees. Sometimes members tend to assume that every coin that a management committee member uses in his private projects is money embezzled from cooperatives. The members therefore become resentful of private projects of the management committee and sometimes even become involved in acts of sabotage. Such misunderstanding or false allegations hurt the officials and demoralize their efforts in running the cooperatives. This sometimes results in the resignation of very competent management staff, thereby damaging the cooperative. Gititu again offers an example of a cooperative characterized by such allegations. This reveals the essence of enlightening members about their rights and responsibilities and those of the management committee. This would minimize conflicts in regard to decisions made by either party.

### *3.2.5 Diversification*

Diversification has been advocated as a useful strategy to buffer the effects of variability of income, preventing net income falling below some minimum level,

**Table 2.4.** Diversification by cooperatives.

| Sub-sector | Performance | Cooperative | Diversification activity |
|---|---|---|---|
| Dairy | Successful | Githunguri | Artificial insemination |
| | | | Foodstuff |
| | | | Rental houses |
| | | | Investment in cooperative union shares |
| | | | Fuelling station |
| | | Limuru | Artificial insemination |
| | | | Sale of food |
| | | | Ploughing services |
| | | | Rental houses |
| | Unsuccessful | Nderi | Undeveloped pieces of land |
| | | Kinale | Sale of food |
| Coffee | Successful | Komothai | None |
| | | Ndumberi | Public address system |
| | | | Letting out idle land |
| | Unsuccessful | Mikari | None |
| | | Gititu | Undeveloped pieces of land |

and increasing the ability of an enterprise to withstand unfavourable conditions (Heady, 1952; Shertz, 1979; Pope and Prescott, 1980; Harsh *et al.*, 1981). The most obvious driving forces for diversification have been short-term effects of risk and instability.

Successful cooperatives had diversified in other income-generating activities in addition to their core business (Table 2.4). The generated income went into running of the cooperatives, the cost of which would have otherwise been covered by deductions from farmers' proceeds. Diversification augmented income for respective cooperatives and this directly or indirectly increased the payout rates or net income paid to members. This built members' confidence in the cooperative and also helped them foster a positive attitude towards the management committee, reducing the chances of wrangles and hostilities. Nevertheless, investment decisions have to be made productively and with the intention of generating additional income for the cooperative. Unsuccessful cooperatives such as Nderi and Gititu made poor investments in plots that were left idle. By sinking funds into non-income-generating investments that are not easily liquidated, these cooperatives find themselves without sufficient liquidity to meet other important financial obligations.

### *3.2.6 Large membership and quantity of produce marketed through cooperatives*

For the most part, the successful cooperatives had a much larger membership than the unsuccessful ones (Fig. 2.4). More members had the advantage of increasing membership share capital and production that benefited the cooperatives through economies of scale. For a cooperative on the verge of collapse, membership share capital may be considered a last-resort source of funds. This was noted with Kinale dairy. Since the cooperative had low sales, which would have translated to lower payout rates than those being offered

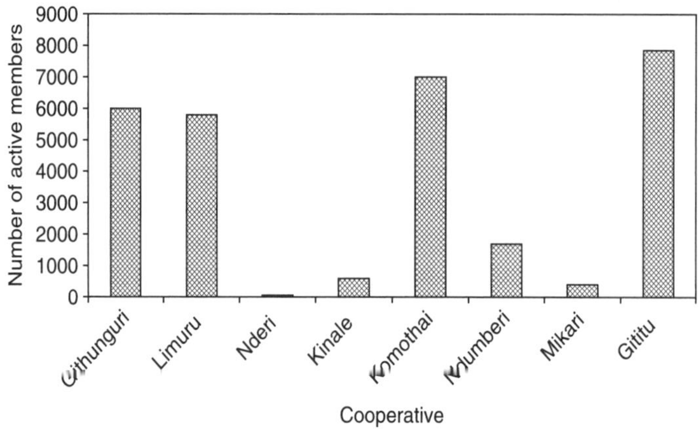


**Fig. 2.4.** Number of active members per cooperative (2004).

by competitors in right of the high fixed costs, it drew some amount from the membership share capital and used it to adjust the rates upwards so as to be at par with competitors (Afrodane Ltd and hawkers). In so doing, it managed to retain farmers who could have otherwise diverted deliveries to the competitors with the intent of benefiting from the attractive rates.

In 2004, Komothai coffee growers' cooperative had 8200 members in total, of which 7000 were active. Likewise, Githunguri dairy cooperative had a high number (6000) of active members. Unsuccessful cooperatives, on the other hand, tended to have low membership. The active members for Kinale were 550 while those for Nderi were at most 50 despite having a registered membership of 1670. The mass withdrawal of its membership is attributed to aggressive hawking with which Nderi was unable to cope. An exception of a cooperative that had a large membership but was doing poorly was Gititu. It had a total membership of 7800, of which 89% were active but performed poorly because of wrangling among members which interfered with normal operations. There was delayed marketing for harvest of the 2003/04 season and members did not receive payment then. Likewise, a splinter from the giant cooperative did not market produce in the 2001/02 season. These interruptions from normal operations negatively affected the average promptness of payment and payout rates, and hence exacerbated the poor performance.

Cooperatives with a large membership benefit from the higher volumes of output that they generate. The resulting higher revenues reduce average fixed and quasi-fixed expenses, thereby increasing potential profit margins and payouts to members. Comparably, the proportion of sales accounted by these expenses tended to be higher with the unsuccessful cooperatives. For instance, in 2003, the expenditure accounted for 52% of sales for Gititu while it was 42% and 32% for Komothai and Ndumberi, respectively. In the dairy sub-sector, the expenses accounted for 55% of sales for Nderi while it was only 14% and 12% for Githunguri and Limuru, respectively. Figure 2.5 presents the volume of milk production for the selected dairy cooperatives. As shown, not only do Githunguri and Limuru produce significantly more milk than Nderi

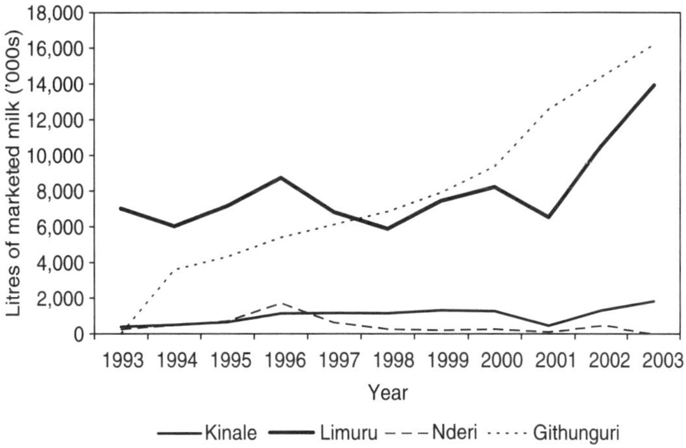


**Fig. 2.5.** Milk production by the cooperatives (1993–2003).

and Kinale, but they also show a general upward trend in output through the years. In contrast, the output trend for Nderi and Kinale is low and largely flat. Much of this can be attributed to the stiff competition from hawkers, who are a particular threat to the smaller cooperatives since they are unable to sustainably pay the competitive rates, largely due to low production that limits them from maximizing benefits from economies of scale.

In the coffee sector, poor payout rates, increased corruption, mismanagement and wrangling among members resulted in a high level of neglect and inefficient production among discontented farmers. In 1996, the threshold acreage under coffee required for a grower to be registered as an individual planter (rather than a smallholder) was reduced from 20 to 5 acres. Together with the other problems facing the coffee sector, the withdrawal of the cooperatives' largest members has led to a considerable fall in total coffee production by cooperatives.

### *3.2.7 Effective record keeping and the provision of credit and farm inputs*

Good record keeping improves transparency and promotes accountability. The availability of financial records detailing the use and allocation of cooperative funds is essential to the effective use of the cooperative finances. Such records impede the wasteful use of funds and constrain rent-seeking opportunities by tracking the use of funds. Such records can also be used to resolve complaints regarding members' payments, the allocation of funds by management committees, etc. Keeping records also provides management with information that can identify leakages that limit the profitability of the cooperative. The benefits to record keeping also increase with computerization, which allows for easier access and analysis of the available information, and increases the speed and accuracy of accounting procedures, thereby making the processing of members' payments more efficient and minimizing delays. Successful cooperatives such as Githunguri and Limuru dairy were computerized.

Most of the successful cooperatives also run their own SACCO. SACCOs play a crucial role by providing liquidity-constrained farmers with credit. This enables

**Table 2.5.** Type of information technology and credit provision strategies.

| Sub-sector | Performance | Cooperative | Computerized | SACCOs | Stores |
|---|---|---|---|---|---|
| Dairy | Successful | Githunguri | Y | Y | Y |
| | | Limuru | Y | N | Y |
| | Unsuccessful | Nderi | N | N | N |
| | | Kinale | N | N | Y |
| Coffee | Successful | Komothai | N | Y | Y |
| | | Ndumberi | N | Y | Y |
| | Unsuccessful | Mikari | N | N | N |
| | | Gititu | N | N | N |

Y = availability, N = unavailability.

farmers to settle their financial obligations while awaiting their cooperative payments. Cooperatives without SACCOs extended credit to farmers principally for farm inputs and not for non-agricultural activities. In such cases, credit fungibility was common as loans meant largely for buying farm inputs were often diverted to other needs. With SACCOs offering an alternative source of finance, credit fungibility as well as illegal sales of coffee cherry at the farm gate for purposes of generating immediate cash was checked. Ndumberi attributed the lack of coffee hawking by its members to the attractive rates paid by the cooperative as well as financial support provided through the cooperative's SACCO.

A common feature of the successful cooperatives is that they all had stores where members could easily access farm inputs, including fertilizers, pesticides, herbicides, animal feeds and farm implements. The inputs are normally supplied to farmers on credit and deducted from their individual payments as necessary. Table 2.5 indicates the availability of credit and input services and the computerization of record keeping across the sample cooperatives.

## 3.3 Failure factors

Beyond the absence or insufficiency of the factors just identified as key correlates of success, we could separately identify several factors most closely associated with unsuccessful cooperatives: debt burden, wrangles and hostilities, and poor management that leaves the cooperative vulnerable to competition from hawkers.

### *3.3.1 Debt burden*

The level of debt that cooperatives faced was a key factor that differentiated the successful cooperatives from the unsuccessful ones. Most of the less successful cooperatives had debt burdens several times greater than their successful counterparts (Fig. 2.6). Much of the debt was accumulated from unpaid development loans. These loans were poorly monitored and in many cases not allocated productively. There were also debts towards farm input suppliers, employees who often go unpaid for months, utility providers and so on.

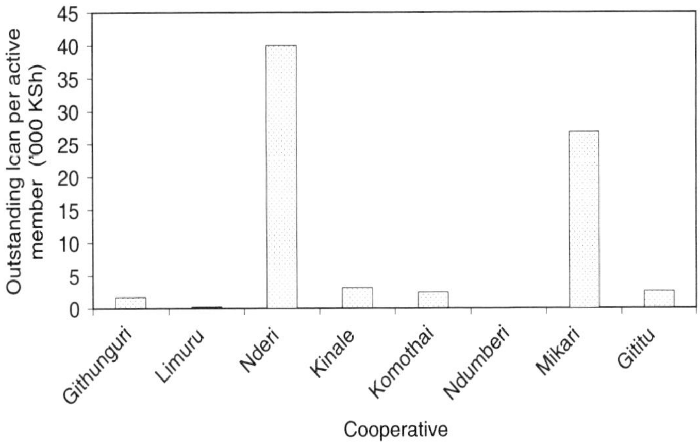


**Fig. 2.6.** Outstanding loan per active member.

As repayment of debts necessarily resulted in lower payouts to farmers, large debt burdens posed a significant problem for cooperatives. Low payments due to debt-servicing deductions made many members sell their output elsewhere to escape shouldering the debt implicitly through lower coffee or dairy unit payouts. Illegal sales of coffee cherry to private pulpers or individual hawkers increased. In the dairy sub-sector, where selling to individual traders was legalized in 1992, discontented dairy farmers could easily sell their milk to hawkers or directly to commercial milk processors. This became a self-reinforcing problem as the drop in the active membership lowered total cooperative output, decreasing gross revenues and increasing average fixed costs, including debt repayment, shouldered by the remaining active members. The result was a vicious cycle.

As a coping strategy to cut costs, some underperforming cooperatives undertook a massive retrenchment of staff. Nderi dairy had 22 employees in 1996, but retained only 3 in 2004; Mikari coffee cooperative had 10 staff in 2002, and only retained 7 in 2004; and Kinale dairy had 20 employees in 2002, but planned to retrench employees soon when we interviewed them in 2004. Similarly, Gititu coffee cooperative retrenched 105 employees. While this reduced the monthly salary expense for Gititu from about KSh 2 million to KSh 600,000, the retrenched employees and their families organized to challenge the management committee's decision. This aggravated the already hostile environment that characterized the cooperative.

### 3.3.2 *Wrangles and hostilities*

The Gititu experience following staff retrenchment offers a good segue to the next correlate of cooperative failure: internal dysfunction. In Gititu, political ambitions by a few self-interested members ignited cooperative-wide wrangles that severely compromised the performance of the cooperative. Irreconcilable feuds caused Gititu not to market coffee in the 2003/04 crop season; consequently, the farmers were not paid for their production. A renegade group of farmers blocked the transport of coffee from factory stores to the millers. In addition, some committee members not successful in securing executive committee positions agitated for a split so as to become leaders of the splinter cooperative. Mounting tensions and stalling production finally culminated in the break-up of the giant Gititu cooperative in January 2003. The splinter society took five factories while the former was left with eight. The split was effectively imposed by renegade leaders in that no liquidator was engaged and a majority of the members voted against the split (1470:541). The process thus seemed to violate most standards of proper procedure. At the time of the split, the giant cooperative had a heavy debt burden for which members of the splinter group refused to accept liability. All the debt was therefore left to be serviced by the remaining members of Gititu coffee cooperative though there was still a complaint on the issue. To settle the controversy, the matter was taken to court. Subsequently, an arbitrator (District Cooperative Officer) was designated to lead in resolving various issues such as modalities of repaying the remaining debts and sharing of common assets between the two cooperatives.

These disagreements and their associated costs contributed to low payout rates and delayed disbursements to farmers. These disenchanted farmers, in turn, neglected coffee or intercropped with heavy crops such as napier grass, lowering coffee quality and thereby hurting not only their own revenues but also those of their fellow cooperative members by bringing down average product quality and thus prices for all in the cooperative. Some farmers even illegally uprooted coffee in favour of added acreage under feed crops in support of dairy farming.

Mikari coffee cooperative was also plagued by wrangles and hostilities among its members with the consequence that no coffee was sold during the 2000 season. The management committee itself was subject to infighting that saw them break into two rival factions, one pro-chairman and the other supporting the vice-chairman. Most farmers sold their coffee to illegal hawkers while the few who delivered to the cooperative had their coffee stored and sold together with the following year's harvest. Mikari financial constraints also made it impossible for them to repair coffee-drying beds and to buy materials such as sisal rolls, which were destroyed by ants, thereby compromising coffee quality.

A weak regulatory environment with limited capacity for effective contract enforcement also makes some cooperatives vulnerable to exploitation by deceitful businessmen. A case in point is Kinale dairy cooperative. In 2000, the cooperative delivered milk on credit to a certain businessman for months only for him to default on his debts at a later date. The cooperative lost KSh 1.9 million in this episode. This generated disunity and distrust among the members of Kinale dairy, some of whom blamed the management committee for incompetence and insufficient oversight and enforcement of the contract. This in turn led to the withdrawal of some members and the entire management committee was voted out, disrupting continuity. Revitalizing the cooperative has since proved very difficult.

### *3.3.3 Vulnerability to competition*

In the liberalized market environment of the Kenyan dairy sub-sector, hawkers have become serious competitors for some cooperatives. Just as predators systematically attack the weak rather than the stronger members of a herd, so has competition from private hawkers been a threat primarily to societies that have been less well managed. Lower fixed costs and a more flexible operating structure allow hawkers to offer more favourable spot market terms, such as prompt payment and better prices. Cooperatives that fail to offer competitive prices and timely payments thus become vulnerable. For instance, while Kinale dairy cooperative paid farmers every 1–3 months, Afrodane, a new processing plant in the area, paid their suppliers weekly. In 2004, Nderi paid its members an average of KSh 18 per litre of milk while hawkers paid KSh 22. Similarly, at Kinale, hawkers offered KSh 17–18 per litre of milk, while the cooperative paid only KSh 14. However, while the hawkers could provide regular payments at higher rates, they could not offer other services, such as credit for farm inputs, veterinary services, animal feed and the like. To benefit from both cooperatives' services as well as prompt payment, some farmers served both markets. Reinforcing our earlier points, those cooperatives who could provide a fuller range of services and whose scale of operation enabled them to reduce average fixed costs and pay farmers a bit more and promptly were better positioned to outcompete these new entrants into the sub-sector.

# 4 Conclusion and Recommendations

This study identified factors associated with the success and failure of agricultural cooperatives in the liberalized Kenyan rural economy. Given the success and failure factors identified, it is important that government, donors and non-governmental organizations (NGOs) implement policies promoting success and inhibiting the failure in order to achieve improved effectiveness and efficiency of agricultural cooperatives and, consequently, to improve the economic welfare of smallholder farmers.

More specifically, the extension of anti-corruption activities in the procurement and investment affairs of cooperatives is likely to check financial mismanagement. Further, value addition, capacity building of management team, provision of credit and improved record keeping, are all important in enhancing the functioning, productivity – and thus the likelihood of success – of the cooperatives as well as farmers' participation and efficiency in input, output and services markets.

Because farmers are the main decision makers in cooperatives, improving their training in, and sensitization to, cooperative management issues is essential. This would promote more informed decision making. However, financial constraints hinder the organizations from carrying out training. Cooperatives might therefore consider contracting with processing and marketing firms to deliver production extension services to farmers as part of a credit-input package. A major challenge arises when farmers decide to sell to other firms rather than to the one that provided the extension services. For this reason, policies geared towards enforcing such contracts are necessary.

Cooperatives and other producer organizations provide a framework for sharing information, coordinating activities, making collective decisions at local level, achieving economies of scale where they exist and amplifying the voices of thousands of otherwise voiceless individuals. They thus play a prospectively valuable role in rural development in Kenya and other low-income, agrarian countries. Helping producers organize and strengthening the capacity of rural producer organizations is one way to combat rural poverty (World Bank, 1995). Therefore, producer organizations are invaluable to the rural developing economy. But there are no guarantees of success.

## Acknowledgements

This study would not have been possible without the financial support provided by the SAGA project through Cornell University, and USAID-Kenya through Tegemeo Agricultural Monitoring and Policy Analysis Project (TAMPA) II. The authors thank officers in the Ministry of Cooperative Development and Marketing, staff of the Coffee Board of Kenya and management committee members of various cooperatives for contributing their time and insights, Wycliffe Oyunga and other Tegemeo research assistants for data assistance, and Chris Barrett, Andrew Mude, Heidi Hogset, John McPeak, John Omiti, Emma Stephens and the scientific staff at Tegemeo Institute of Agricultural Policy and Development for comments on earlier drafts.

## References

Cook, M. (1997) *Organizational Structure and Globalization: The Case of User Oriented Firms*. In: Nilsson, J. and Van Dijk, G. (eds) *Strategies and Structures in the Agro-food Industries*. Van Gorcum, Assen, The Netherlands.

DCO (District Cooperative Office) (2003) *Kiambu District 2003 Annual Report*. Ministry of Cooperative Development and Marketing, Kiambu.

Harsh, S.B., Conner, L.J. and Schwab, G.D. (1981) *Managing the Farm Business*. Prentice-Hall, Englewood Cliffs, New Jersey.

Heady, E. (1952) Diversification in resource allocation and minimization of income variability. *Journal of Farm Economics* 34, 482–496.

ICA (International Cooperative Alliance) (1995) Farmers Organisations and Rural Cooperatives. Geneva.

ICO (International Coffee Organization) (2003) *Coffee Statistics*. London.

IEA (Institute of Economic Affairs) (2002) *The Little Fact Book: The Socio-economic and Political Profiles of Kenya's Districts*. Nairobi.

Ministry of Cooperative Development and Marketing (2003) *Kiambu District 2003 Annual Report*. Republic of Kenya.

Pope, R.D. and Prescott, R. (1980) Diversification in relation to farm size and other socioeconomic characteristics. *American Journal of Economics* 62, 554–559.

Republic of Kenya (2002) *National Development Plan 2002–2008*. Ministry of Planning and National Development, Nairobi.

Shertz, L.P. (1979) The major forces. In: Shertz, L.P. (ed.) *Another Revolution in US Farming*. USDA, Washington, DC.

USDA (United States Department of Agriculture) (1997) *Cooperatives in International Trade*. USDA, Washington, DC.

World Bank (1995) *Empowering Farmers in Sub-Saharan Africa: Best Practices*. Washington, DC.

# 3 Institutional Incompatibility and Deregulation: Explaining the Dismal Performance of Kenya's Coffee Cooperatives

A.G. Mude

## 1 Introduction

Promoting the capacity of the poor to mobilize for collective action is increasingly championed as an effective mechanism to empower the poor and amplify their ability to leverage resources for their benefit. The increasing attention paid to the catalytic role that membership-based organizations of the poor can play in improving the welfare of the poor is a natural and complementary extension to the recent emphasis on decentralization of development services and on the positive role social capital can play to stimulate rural, traditional economies.

While a general consensus exists that empowering the poor to take a proactive role in their development should be a central pillar of development efforts, it is not as clear that membership-based organizations are always the most effective means to achieve such ends or that they necessarily improve the welfare of their members. Numerous studies have documented cases in which collective organizations with a development mandate have failed to meet their stated objective, at times even leaving members worse off (Morduch, 1999; Rahman, 1999; Gugerty and Kremer, 2004). Drawing from such experiences, a nascent literature now studies how the very act of creating a membership-based organization can give rise to incentives that work against the original intended goals of the organization (Howes, 1997; Stiles, 2002; Gugerty and Kremer, 2004).

This chapter highlights this issue from the perspective of the smallholder coffee sub-sector in Kenya. It investigates the role that market decentralization within the coffee sector in Kenya played in stagnating smallholder coffee cooperatives. Divestment, also known as market decentralization, describes the form of decentralization whereby government responsibilities and authority in planning and organizing economic activity is transferred to non-public institutions with the intention of increased efficiency of production, marketing and local involvement.

Before the advent of structural reforms in the 1990s, coffee was Kenya's top foreign exchange earner. At that time the government played a key role in regulating


(eds C.B. Barrett, A.G. Mude and J.M. Omiti)

the activities of the nation's coffee cooperatives. Liberalization brought with it a withdrawal of government involvement in cooperatives and a series of reforms which culminated in the new Cooperatives Act of 1998. By this act of divestment, the government gave up its policy-making jurisdiction over the economic activities of cooperatives. Grower-members have fully owned and managed their cooperatives ever since. Payments made to coffee growers subsequently plummeted and the coffee smallholder industry found itself mired in increasing levels of corruption, political opportunism and gross mismanagement.

In a similar vein to Gugerty and Kremer (2004), we argue that the regulation change that made members fully responsible for running cooperatives changed the incentive structure facing members and officials in a way that counteracted the expected benefits to divestment. More specifically, we contend that unregulated access to cooperatives' coffers now enjoyed by elected cooperative officials significantly increased the rents that self-interested officials could expect to extract from the treasury. With limited government oversight, the socio-economic and institutional matrix upon which divestment was introduced facilitated the corruptibility of elected cooperative officials. The lack of a formal regulatory structure with credible enforcement mechanisms, the presence of informal electoral practices conducive to vote-buying and the legal support for local monopsonies that facilitates exploitative pricing all served to increase the incentives for rent-seeking. Consequently, and as widespread anecdotal evidence confirms, the process of electing a governing board from among the membership became subject to manipulation by rent-seeking candidates.

Based on this premise, we build our case by invoking a model of election capture, whereby we argue that in the absence of formal, credibly punishable rules against bribery, members of a cooperative can rationally elect an official in open democratic elections who they know will reduce the returns they would otherwise receive from the cooperative.[1]

The very nature of corruption, however, makes it a tricky task to collect the requisite data accurately so as to undertake a robust empirical test that captures the presence of rent-seeking activity and its effect on the welfare of members. As such, we take a somewhat indirect approach to investigate the effect of cooperative mismanagement on farmer productivity and welfare. Drawing on evidence from the smallholder coffee sub-sector in Murang'a district, Kenya, we show that attributes of cooperatives indicative of rent-seeking activity, such as performance measures and members' perceptions, are statistically significantly associated with reduced farm-level technical efficiency. The implication is that changes to the features of the institutional environment that give rise to rent-seeking behaviour could reduce corruption and increase productivity and rural incomes. The lesson we seek to high-

[1] A theoretical companion piece to this chapter models the susceptibility of cooperatives to capture by corrupt and opportunistic members who would extract rents from collective earnings for their own benefit. We show that under institutional circumstances similar to those that underlie coffee cooperatives in Kenya, corruption can very easily take root among the leadership. The model implies that larger cooperatives are more likely to be captured by corrupt individuals and that rent-seeking results in a drop in payments to farmers and production. Although the model is briefly summarized in Section 3.2, interested readers can contact the author for details on the theoretical model.

light in this chapter is that while market decentralization may empower producer organizations, a conducive institutional arrangement as well as a politically mature and well-informed membership is necessary to tap the potential of divestment.

The rest of the chapter is structured as follows. Section 2 gives a brief description of the institutional environment within which coffee cooperatives function and provides the contextual justification for the argument we build. Section 3 focuses on the likelihood of election capture by rent-seeking individuals. We highlight the key institutional features facilitating rent-seeking and summarize a model of election capture built within an institutional apparatus similar to that underlying the coffee cooperatives in Kenya. Section 4 introduces the data and offers empirical evidence to support our claims. A discussion of the findings concludes the chapter in Section 5.

## 2 Coffee Cooperatives in Kenya

### 2.1 Organizational structure

Since its introduction as a cash crop in the early twentieth century, coffee has traditionally been the backbone of Kenya's rural highlands economy. Coffee was the nation's top foreign exchange earner from independence in 1963 until it was surpassed by tourism in 1989. Since then, national coffee earnings have steadily declined and currently rank fourth after tourism, tea and horticulture (Karanja, 2002). From its heyday in the 1970s and 1980s, when international coffee prices were high and the government regulated production and marketing systems, Kenya's coffee sector now finds itself in crisis.

Small-scale production systems dominate Kenya's agricultural sector, accounting for approximately 60% of marketed output. The coffee sector is particularly affected by the fragmented nature of small-scale production, with over 75% of the land under production controlled by smallholder farmers.[2] For this reason, the smallholder coffee sector has traditionally been organized into cooperatives in order to facilitate regulation and improve the effectiveness and efficiency of coffee production, marketing and the provision of key inputs such as fertilizers, pesticides, credit and extension services.

The socio-economic landscape at the time, combined with the physiology of coffee, initially justified such an organizational structure. As soon as coffee cherries are harvested they begin to ferment, a process that affects the quality of the bean. This requires the cherries to be pulped as soon as possible after they are picked. Pulping stations are thus needed to be sufficiently close to the farmer to assure prompt delivery. As pulping stations involve large set-up costs that are prohibitive to individual small growers, and as one pulping station can cater to hundreds of growers, it made sense to pool their resources together within a cooperative framework to share the set-up costs and enjoy the resulting local economies of scale. These pulping stations are called 'factories'. A single cooperative can encompass multiple factories.

[2] Any farmer with less than 5 acres of land under coffee production is classified as a smallholder-grower.

All smallholders were, and continue to be, legally bound to market their coffee through cooperatives. Each cooperative controls a loosely defined catchment area.[3] All coffee growers whose land falls within a particular catchment must register as a member with their respective cooperative. Conditional on cooperative membership, the choice of which factory to deliver one's coffee to is more flexible, being a function of proximity and political considerations.[4] Essentially, the farmers' only task is to grow the coffee, and deliver their product (coffee cherry) to their chosen pulping station.

In return, growers are paid twice annually. The first payment comes at the beginning of the season (preharvest) as part of what is called coffee advance payments (CAPS). CAPS are often calculated as the lowest expected payment per kilogram for the coming season. At the end of the season, after coffee sales have been realized and all requisite deductions taken, the remaining amount is then distributed to farmers as the second payment.

At the factory, the cherries from each grower are weighed, recorded and then put through the initial phase of processing. Primary processing involves the sorting and pulping (removing the coffee bean from its outer fruit) of the cherry. The beans are then laid out to dry and thereafter stored in the form of 'parchment coffee', awaiting transportation to millers for secondary processing. Millers hull and clean the parchment coffee to produce what is known as green (raw) coffee. They then sort and grade the coffee by size and quality, bag it according to expected value, and send it forward to the coffee auctions through which, by law, all Kenyan coffee must be sold. The coffee in each bag belongs to one particular factory within the cooperative. This allows for interfactory intracooperative price variation, which takes into account variation in quality, and thereby reduces the potential for moral hazard at the larger cooperative level.[5] The transaction between buyer and seller at the auction is often carried out on behalf of the cooperative by an agent hired by the miller.

Once the coffee is sold, the miller deducts his share of the commission and sends the rest to the cooperative. The larger cooperative management then deducts all of its operating costs including loan repayments, services and maintenance expenses, and other fees. The deductions are made from factory kitties in proportion to each factory's membership to the total (i.e. uniform deduction per cooperative member). The remaining funds are distributed to factory managers who further deduct the costs of factory-level operations, then distribute the remaining money to farmers as their second annual payment.

[3] By loosely defined, we mean that no legally defined boundaries enclose the catchments. A combination of natural borders (rivers, main roads, etc.), political boundaries (location, sublocation, etc.) and initial distribution of households across cooperatives at the time of their conception forms the basis of generally accepted informal boundaries that define catchments.

[4] While in theory, one is free to market through one's factory of choice, switching pulping stations is generally frowned upon and is likely to invite sociopolitical retribution that limits one's practical choices.

[5] Each factory pools its coffee cherry together such that each farmer's contribution is indistinguishable. Individual growers may therefore have the incentive to shirk on efforts to produce high-quality coffee and freeride on the efforts of others within their factory group. Factory-specific pricing limits freeriding to the factory level.

## 2.2 Institutional evolution: the reform period

Reacting to pressure from international donors in the late 1980s and early 1990s, the government enacted a series of reforms aimed at the eventual liberalization of the Kenyan economy. As Kenya's main foreign exchange earner, the coffee sector was a major target for reforms. In October 1992, the Coffee Board of Kenya (CBK) was authorized to conduct the Nairobi coffee auction in dollars and was later granted the permission to also pay farmers in dollars. This lifted the implicit tax burden, estimated at 29%, which farmers previously faced due to an overvalued Kenyan shilling (Ephanto, 2003). Changes were also made to improve the system by which growers received revenues generated from sales of their coffee. The new system eschewed the previous practice of pooling payments across time and farmers in order to reduce price variation across time and price spread between growers. Under the new 'direct payments' scheme, growers could now receive payments for sales of their own coffee on a weekly basis, thus improving their liquidity and stimulating incentives for the production of quality coffee.[6]

Further reforms included the licensing of four new commercial millers in 1993, thereby dismantling the monopoly in the milling sector previously held by the Kenya Planters Cooperative Union (KPCU). In 1996, efforts to stimulate production changed the rules governing the licensing of coffee growers. The threshold acreage under coffee required for a grower to be registered as an individual planter (rather than a smallholder) was reduced from 20 to 5 acres. The idea was to improve incentives for higher quality by increasing the number of growers benefiting from the 'direct payments' scheme.

The policy change of particular interest to this chapter involves the new Cooperatives Act of 1998 which gave farmers complete autonomy over the activities of the cooperative. Prior to 1998, the government played a major role in the running of cooperatives through the office of the commissioner of cooperatives, and their field-agents, led by district cooperative officers (DCOs). Although members owned the cooperatives and elected their board members, the commissioner's office had powers to dissolve the governing board, call for fresh elections or directly appoint a caretaker committee. DCOs were also involved in the vetting of candidates for management positions. More importantly, DCOs were counted as extra-official members of the governing board and were mandatory signatories to all cheques and withdrawals made by the management. The commissioner's office was also the sole agent authorized to audit society accounts.

Under the new policy, government no longer had any policy-making jurisdiction over the economic activities of cooperatives and took on a minimal advisory role. Under the Cooperatives Act of 1998, cooperative board members were free to conduct elections as they pleased, make hiring decisions of their choice and contract, if they so wished, their auditors of choice. DCOs were no longer required to co-sign on any financial transactions. Extension services, previously provided through the Ministry of Agriculture, were also withdrawn. In effect, cooperative board members now had complete authority

[6] The benefits to this policy did not accrue directly to smallholder farmers as the pooling of their payments continued at the cooperative level.

over the running of cooperatives without the oversight of a regulatory agency with any teeth to prosecute malfeasance. These changes increased the incentives to rent-seeking by providing board members with unfettered access to cooperative coffers without the fear of prosecution.

Thereafter, payments to growers plummeted amid growing political opportunism at the grass-roots level that damaged farmer morale and raised the level of corruption and mismanagement in cooperative administration. Through the years, a general decline in the attendance of annual general meetings (AGMs), called to elect board members, ensued as growers became disillusioned with the electoral process. Widespread belief has it that a majority of those who continued to attend AGMs are bribed for a pittance. Indeed several of the growers we talked to unabashedly acknowledged that they had indeed accepted bribes of Ksh 100 (roughly US$1.40) or offerings of the local brew on Election Day to vote a particular candidate. In this way, corrupt board members entrench themselves and embezzle the proceeds of coffee sales, further eroding the ailing cooperatives.

## 2.3 Weak international markets are not the key factor

In this chapter, we argue that much of the deterioration of the coffee cooperatives can be attributed to divestment of cooperative management that culminated in complete unregulated member autonomy over all aspects of cooperative activities. Anticipating critics who may instead argue that declining international coffee prices explain the bulk of Kenya's smallholder coffee crisis, we provide some evidence that other key coffee-exporting countries, similarly affected by low global coffee prices, were not as adversely affected as Kenya. This difference, we claim, is the result of the poor sequencing, design and implementation of institutional reform in Kenya's coffee sector.

Figures 3.1–3.3 compare various indicators of the performance of Kenya's coffee industry against five other coffee-exporting nations that also principally grow high-quality Arabica beans. For ease of comparison, we standardize all 1994 figures to parity at one hundred, and examine the relative trend of changes in the indicators vis-à-vis their 1994 position. The price data were deflated to the 1994 level to control for inflation. To better inform the comparisons, Table 3.1 presents the 1994 levels data for the selected indicators (yield, production and price) across the six countries.

While Kenya posts the lowest total production and average yield of the six countries, it received the highest price for its coffee exports, a testament to the renowned premium quality of its coffee. Peru on the other hand, second only to Kenya in its low yield and production, also posted the lowest price. Of the six nations, Ethiopia had the highest total production in 1994, almost three times more than that of Kenya. Costa Rica, for its part, posted significantly higher yields than the rest.

Figure 3.1 compares relative changes in the average price paid for green coffee across the chosen countries. There has been a general downward trend in prices and all nations received a lower return for their coffee in 2002 than

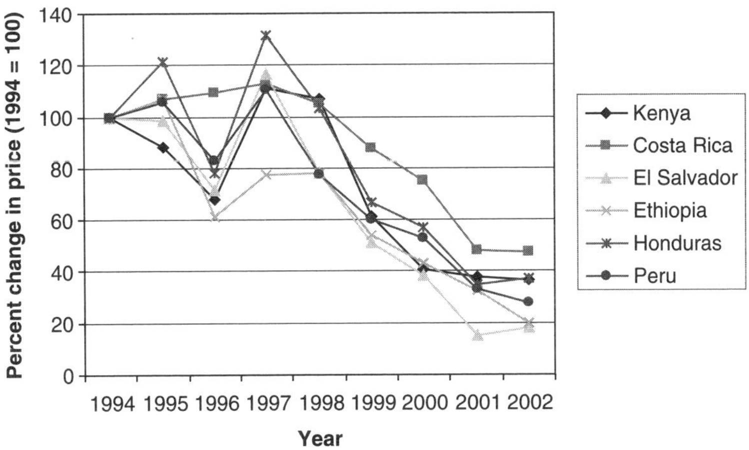


**Fig. 3.1.** Green coffee prices for select exporting nations. (From International Coffee Organization (ICO) Statistics.)

in 1994. Kenya's coffee in 2002 fetched approximately 35% of the price that it did in 1994. Costa Rica and Honduras, whose 2002 coffee prices were approximately 47% and 37% of their 1994 level respectively, were the only countries whose coffee prices fell less than Kenya's. Both Ethiopia and El Salvador witnessed close to 80% deterioration in the real price paid for coffee between 1994 and 2002, while Peru faced a 70% drop.

Despite the steep price decrease, Peru posted an impressive growth in total production of close to a 100% between 1994 and 2003 (Fig. 3.2). Kenya on the other hand, which had relatively modest price decreases, witnessed a decrease in output to 65% of its 1994 level in 2002. This significant contraction in output was mirrored only by El Salvador, which also posted the largest decrease in its prices.

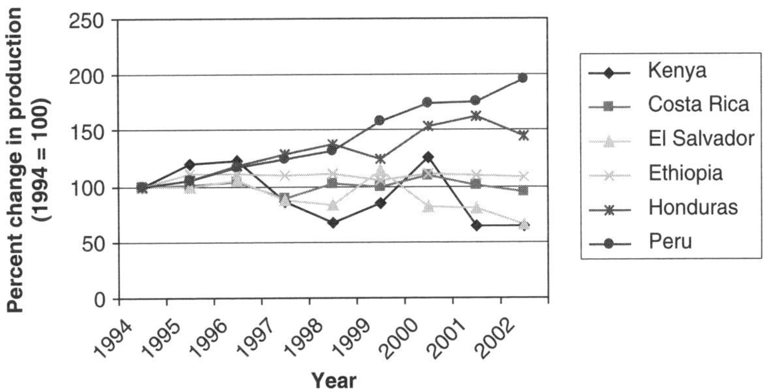


**Fig. 3.2.** Coffee production for select exporting nations. (From Food and Agriculture Organization, FAOSTAT.)

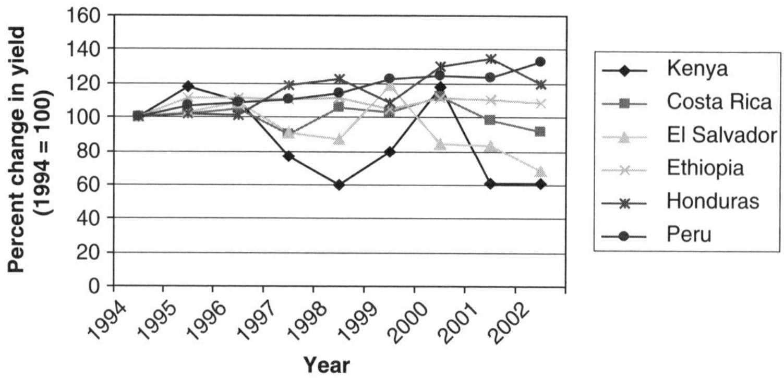


**Fig. 3.3.** Coffee yields from select exporting nations. (From FAOSTAT.)

Changes in average yields, another indicator of sector performance, further highlights the stagnation of Kenya's coffee industry. Despite posting the lowest relative yields in 1994 (see Table 3.1), it also posted the largest decrease in yields (Fig. 3.3). In 2002, Kenya's coffee yield was estimated to be only 60% of its 1994 level. Meanwhile, Peru and Ethiopia, both experiencing a relatively larger decline in price than Kenya, posted gradual and consistent increases in yield.

These data indeed suggest that declining world prices for coffee cannot fully explain the collapse of the Kenyan coffee industry. Other nations, similarly or worse affected by low prices, were not only able to maintain their aggregate output and average yields, but were in most cases also able to increase yields and boost production. This indicates that additional circumstances specific to the Kenyan coffee industry must have amplified the negative consequences of the unfavourable international market.

Moreover, these figures are national aggregates and therefore do not present an adequate picture of the plight of the smallholder sub-sector. In the 1990s, the smallholder share of national coffee output was just below 60% on average. However, while production in the plantation sector decreased by around 39% in the last decade, the decline in smallholder farms was around 47% during the same period (Karanja and Nyoro, 2002). Furthermore, smallholder yields are far less than estate yields. Table 3.2 presents a comparison of smallholder and estate production and yields through the 1990s.

For the smallholder coffee farmer in Kenya, the combination of these three trends – declining prices, output and yields – contributed to a significant

**Table 3.1.** 1994 Levels for coffee indicators of select exporting countries.

| | Kenya | Costa Rica | El Salvador | Ethiopia | Honduras | Peru |
|---|---|---|---|---|---|---|
| [a]Yield (hg/ha) | 5,035 | 13,582 | 8,344 | 8,280 | 7,050 | 5,562 |
| [a]Production (t) | 79,900 | 147,998 | 140,534 | 207,000 | 126,182 | 91,340 |
| [b]Price (US cents/lb) | 251.4 | 137.2 | 194.6 | 184.3 | 163.1 | 97.9 |

[a]From Food and Agriculture Organization.
[b]From International Coffee Organization.

**Table 3.2.** Clean coffee production and yields in Kenya (1991/92–2000/01). (From Karanja and Nyoro, 2002.)

| | Production (t) | | | Yield (kg/ha) | | |
|---|---|---|---|---|---|---|
| Year | Estates | Smallholder | National | Estates | Smallholder | National |
| 1991/92 | 37,520 | 51,977 (58%) | 89,497 | 987 | 439 | 565 |
| 1992/93 | 32,781 | 42,426 (56%) | 75,207 | 859 | 352 | 474 |
| 1993/94 | 33,037 | 39,747 (54%) | 73,516 | 860 | 324 | 457 |
| 1994/95 | 32,795 | 62,567 (65%) | 95,806 | 855 | 510 | 595 |
| 1995/96 | 40,109 | 56,881 (58%) | 97,576 | 1,045 | 464 | 606 |
| 1996/97 | 29,737 | 38,261 (56%) | 67,997 | 748 | 312 | 419 |
| 1997/98 | 22,061 | 32,981 (60%) | 55,042 | 555 | 269 | 339 |
| 1998/99 | 28,700 | 39,400 (58%) | 68,100 | 684 | 307 | 400 |
| 1999/00 | 38,500 | 62,200 (62%) | 100,700 | 916 | 485 | 592 |
| 2000/01 | 26,900 | 24,800 (48%) | 51,700 | 640 | 193 | 304 |
| **Average** | **32,214** | **45,124 (58%)** | **77,514** | **815** | **365** | **475** |

deterioration in the welfare of growers, who for decades had depended on coffee incomes as a secure and relatively lucrative livelihood.

# 3 Election Capture in Kenya's Coffee Cooperatives

## 3.1 Institutional features facilitating election capture

'Election capture' describes the process by which self-interested and corrupt candidates illegitimately manipulate the electoral process in order to secure their victory. Along with the Cooperatives Act of 1998 that increased the expected returns to rent-seeking and decreased the risk of detection and censure, certain features of the existing institutional environment contributed to election capture in Kenya's coffee cooperatives.

### *3.1.1 Perfect vote signalling*

All of the nine coffee cooperatives surveyed for this study conducted their elections in the traditional fashion of *mlolongo*. *Mlolongo*, literally translated as 'line-up', describes the method of having voters line up behind their preferred candidate with the candidate having the longest line winning the election. Clearly, the consequence of such a method is that everybody knows who everybody else voted for. This facilitates vote-buying by offering a free and perfect enforcement mechanism for candidates. A voter who might otherwise simply accept the bribe and thereafter vote independently under a secret ballot regime must now consider the cost of near-certain punishment should he or she deviate. A secret ballot system for democratic majority-rules elections weakly dominates a perfect signalling *mlolongo* approach. *Mlolongo* provides just the enforcement mechanism a rent-seeking candidate could use to advance his or her objective, undermining grower productivity in the process.

### *3.1.2 Local monopsony power*

Kenyan law requires all coffee growers with less than 5 acres of land under coffee to market their output solely through cooperatives. Furthermore, due to poor transportation infrastructure and the need to pulp coffee cherry soon after it is picked in order to avoid quality-reducing fermentation, each cooperative has a legally defined catchment area. Making it illegal for growers to sell their coffee to other potential buyers effectively grants cooperatives local monopsony protection and shields them from potential competition. The logic of organizing to attain an input–output mix at the bottom of the long-term average cost curve assumes a competitive market that requires collective cooperation among small producers who intend to be competitive. Yet, protecting such organizations against competition discourages them from being efficient as there are no longer constraints that force them to maximize the benefits to cooperation. The very motivation for organization, to attain optimal scale in the face of competition, loses its salience under monopsony. In addition, local monopsony protection empowers rent-seeking managers to exploit their growers by forcing them to accept payments lower than the equilibrium price that would obtain under a competitive market.

## 3.2 Modelling election capture in Kenya's coffee cooperatives

A companion paper develops an analytical framework to model likelihood of election capture within Kenya's coffee cooperatives. We base the model on an *n*-person prisoners' dilemma game to investigate the susceptibility of cooperatives to capture by corrupt and opportunistic members who would extract rents from cooperative coffers for their own benefit. Mirroring the institutional environment underlying Kenya's coffee cooperatives, we assume local monopsony power, an unregulated electoral process conducive to vote-buying practices and sufficient returns to rent-seeking. We show that in the absence of formal, credibly punishable rules against bribery, members of a cooperative can rationally elect an official in open democratic elections who they know will reduce the returns they would otherwise receive from the cooperative. Briefly, the model works as follows.

Due to the large rents a corrupt candidate can expect upon winning office, he or she may be willing to buy the minimum required votes to guarantee an election victory. As an honest candidate will not siphon cooperative profits for personal gain, he or she cannot afford to exchange favours for votes and is thus at a competitive disadvantage. A voter, on the other hand, will accept any bribe that is at least equal to the expected loss in welfare that would be sustained *if* he or she were to be the pivotal voter and swing the election outcome. As the number of voters increases, the likelihood that any one voter would be the pivotal voter decreases, making it increasingly cheaper to buy votes and therefore more likely that a rent-seeking candidate will buy his way into office. There is thus an electoral diseconomy of scale at the cooperative level.

The model's implications correspond well with the general trends that characterize coffee cooperatives in Kenya. We find that under fairly general circumstances corruption can very easily take root among the leadership. Indeed, plenty of anecdotal evidence indicates that rent-seeking in cooperatives is widespread. Because vote-buying is easier as the number of voters increases, the model suggests that larger cooperatives are more prone to election capture. Furthermore, as rent-seekers divert some of the revenues intended for distribution among the membership to their personal accounts, the model implies that, *ceteris paribus*, cooperatives infiltrated by a corrupt governing board pay their members less who, in turn, produce less output. As we now show, this is precisely what has happened in the coffee sector in Murang'a district.

# 4 Data Analysis

## 4.1 Data

The data were collected over a 3-month period between November 2003 and February 2004. The effort was part of a larger study of the institutional arrangements in the smallholder coffee, tea and dairy sub-sectors in Murang'a district of central Kenya, a high potential agricultural area on the eastern slopes of the Aberdare ranges endowed with good soils and favourable rainfall.

In order to capture variation at the institutional level, our sampling method was stratified. We first identified coffee cooperatives in Murang'a, collected institutional data at this level and then randomly sampled their members for further in-depth, farm-level surveys. We picked 9 out of 19 coffee cooperatives in the district, purposively selecting the cooperatives so as to achieve the greatest variation in spatial coverage of Murang'a, cooperative size, and subjective performance based on information from the District Cooperative Officer, the District Agricultural Officer and recent payments offered to members for their output. Once the cooperatives were selected, we randomly picked a factory (or more for the larger cooperatives) and from that randomly selected our household sample from the register of members.[7]

At the institutional level, our collection efforts were hampered by a lack of available information, secrecy on the part of the managers and a general unwillingness to provide the information sought. While we had initially planned to collect information on prices paid to farmers, costs of operation, mark-ups, prices received from auctions, services offered to farmers, fees charged for services, structure and duration of contracts, feasible number of prospective contractors, etc. for the past couple of years, we were unable to get a full set of responses from most cooperative managers. Thankfully, the farm-level surveys were relatively successful and we recovered much useful data. We recorded information on inputs and costs of production, credit access and use, farm size, area under coffee, other farm enterprises, social organizations involved, extension services received, and fertilizer use among others. Table 3.3 presents some general statistics about the cooperatives.

[7] Depending on their size, cooperatives in our sample had 1–12 factories. Given the difficulty in accessing selected members in the rainy season of Kenya's hilliest and most mudslide-prone district over seasonal roads, it became too restrictive to randomly survey at the cooperative level as opposed to the factory level. Furthermore, the per cooperative sample sizes would have been too small to capture any interfactory differences. To cater to the possible bias of selecting respondents from one or two factories, we collected basic statistics from each factory.

**Table 3.3.** Descriptive statistics for sampled cooperatives.

| Cooperative name | Number of members sampled | Number of factories owned | Total number of members | Average yield (kg/tree) | Payment (KSh) per kg of cherry | Coefficient of variation of pay across factories |
|---|---|---|---|---|---|---|
| Kamacharia | 18 | 4 | 3760 | 1.94 | 4.82 | 0.43 |
| Gaturi | 48 | 5 | 3752 | 1.97 | 3.97 | 0.80 |
| Weithaga | 20 | 4 | 2101 | 2.51 | 3.27 | 0.22 |
| Kanyenyaini | 15 | 2 | 1249 | 2.96 | 7.68 | 0.18 |
| Kahuhia | 50 | 6 | 3704 | 1.35 | 1.41 | 0 |
| Iyego | 36 | 12 | 7000 | 2.33 | 5.27 | 6.35 |
| Kiru | 19 | 4 | 2837 | 1.12 | 4.51 | 0.16 |
| Kangunu | 21 | 1 | 1320 | 3.54 | 15.85 | – |
| Kiriti | 20 | 3 | 2085 | 1.66 | 7.99 | 0.10 |

As Table 3.3 shows, there is considerable variation among the cooperatives along every dimension. Cooperative size, as defined by the total number of active members, varies from Kanyenyaini with 1249 members to Iyego with roughly 7000. Kangunu, which is only marginally bigger than Kanyenyaini and is the only society with a single factory, also has the highest sample average yield of 2.99 kg of cherry per tree, more than five times that of Kamacharia, which trails in coffee yield at a mere 0.56.

The largest variation, however, occurs in the payments farmers receive from their respective factories. Kangunu, which paid its members KSh 15.85 in the 2002/03 season, paid more than ten times the amount that Kahuhia growers received. The payments, collected from cooperative records, are given as averages across all factories belonging to the cooperative. All members are paid at the factory level with payments varying among factories within any one cooperative due to quality and quantity of coffee produced. The standard deviation of payments across factories, normalized by their respective means, is given in the final column. Note that Iyego posts a higher normalized standard deviation than its mean, suggesting that some members must have been receiving negative effective payments. This is indeed true for several factories where a negative effective payment simply means that not only did the farmers attached to those factories receive no payments for the 2002/03 season, but they were additionally burdened with debt that carried forward into the next season.

## 4.2 Empirical strategy

Our goal is to test for the presence of rent-seeking behaviour in cooperatives and to show that, to the extent it exits, it has an inverse relationship with farm-level technical efficiency. Diminishing efficiency, rather than diminishing production, is a stricter measure of the negative consequence of rent-seeking as it goes beyond merely asking if total output or yields have decreased – a trend that has already been shown to exist in the aggregate. A decline in output itself is not necessarily a signal of weak performance and could simply suggest that other livelihood options or alternatives began to yield higher returns and members were shifting their resources accordingly. Falling output may thus reflect a rational shift in aggregate production patterns in response to changes in the expected returns of available livelihood options. As such, we seek to identify a statistically significant association between farm-level technical inefficiency and corruption or mismanagement at the cooperative level.

The idea is that beyond declines in output, rent-seeking at the cooperative level will result in higher deductions from members' payments, and a reduction in the provision of services generating disincentives for members that will manifest in lower technical efficiency. Growers associated with more corrupt cooperatives are likely to become disenchanted and reduce the effort they apply to the production of coffee. By controlling for observable inputs to production, the resulting measures of technical efficiency proxies, to some extent, for the unobservable level of effort.[8] If measures of cooperative corruption and mismanagement are associated with low levels of technical efficiency, it is possible to attribute part of the poor performance of cooperatives to infiltration by rent-seeking board members.

To tease this out from our data, we conducted three separate but interrelated tests. First, we estimated a stochastic production frontier for coffee yield and used the results to generate a farmer-specific measure of technical (in)efficiency. We then conducted two separate factor analyses to extract proxies that together indicate the likelihood and extent that the various cooperatives are involved with rent-seeking behaviour. The third test uses the efficiency measures generated from the frontier estimation as the dependent variable in an OLS regression aimed at determining the sources of technical (in)efficiency. The rent-seeking proxies generated from the factor analyses are used as independent variables in an effort to gauge the relationship between cooperative-level rent-seeking and individual member technical efficiency.

## 4.3 Estimating technical efficiency

To investigate patterns of farm-level technical efficiency, we estimate a stochastic coffee production frontier and then calculate each unit of observation's deviation from this benchmark of optimal efficiency. Stochastic production frontier models were introduced independently by Aigner *et al.* (1977) and Meeusen and van den Broeck (1977). They provide estimators for the parameters of a linear model with a disturbance that is assumed to be generated from two separate processes: one having a strictly non-negative distribution that parameterizes the inefficiency error term, and the other with a symmetric distribution to

[8] Barrett *et al.* (2006) follow similar logic to show how massive currency devaluation in the Ivory Coast resulted in a reduction in mean-plot-level technical efficiency in rice production attributable to disenfranchisement among farmers.

capture random error. Equation 3.1 briefly summarizes the stochastic production frontier problem.[9]

Suppose firm $i$'s expected output $q_i$ is given by

$$q_i = f(\mathbf{z}_i, \beta)\xi_i \exp(v_i) \tag{3.1}$$

where $f(\mathbf{z}_i, \beta)$ denotes the production function, $\xi_i \in [0,1]$ captures the degree of firm $i$'s technical efficiency where $\xi_i = 1$ signifies optimal production and $\exp(v_i)$ denotes random shocks to which firm $i$ could be exposed. Taking the natural log of Equation 3.1 on both sides yields:

$$\ln(q_i) = \ln\lfloor f(\mathbf{z}_i, \beta)\rfloor + \ln(\xi_i) + v_i \tag{3.2}$$

Defining $u_i = -\ln(\xi_i)$, assuming $k$ inputs we estimate the following production frontier:

$$\ln(q_i) = \beta_0 + \sum_{j=1}^{k} \beta_j \ln(z_{ji}) + \sum_{j=1}^{k} \gamma_i \left[\ln(z_{ji})\right]^2 + v_i - u_i \tag{3.3}$$

Since $u_i$ is subtracted from $\ln(\xi_i)$, restricting $u_i \geq 0$ implies that $\xi_i \in (0,1]$ as specified above.

In estimating the frontier, we assume a normal distribution for the symmetric disturbance term and a half-normal for the distribution of the inefficiency term. Table 3.4 presents summary statistics for the variables used in the estimation.

Our measure of technical efficiency in coffee production is yield in kilograms of coffee cherry per tree. In order to minimize the influence of outliers,

**Table 3.4.** Descriptive statistics for frontier estimation variables.

| Variable | Mean | Standard deviation | Minimum | Maximum |
|---|---|---|---|---|
| Coffee yield[a] | 2.04 | 1.96 | 0.16 | 7.5 |
| Plot area[a] (acres) | 0.56 | 0.49 | 0.00 | 3.00 |
| Land area[a] (acres) | 1.84 | 1.77 | 0.20 | 13.00 |
| Household labour[a] (days) | 44.84 | 38.06 | 0 | 120 |
| Hired labour[a] (days) | 29.37 | 38.35 | 0 | 112 |
| Age of coffee tree[a] | 30.88 | 11.13 | 2 | 50 |
| Preharvest damage[a] | 26.73 | 32.29 | 0 | 95 |
| Inorganic fertilizer (0 = N, 1 = Y) | 0.234 | 0.424 | 0 | 1 |
| Organic fertilizer (0 = N, 1 = Y) | 0.447 | 0.498 | 0 | 1 |

[a]Denotes variables used in natural log form in estimation. For these variables, we followed the common practice of substituting 0.001 for zero-valued observations to log-transformations to be defined across the variables range. Statistics presented for non-transformed variable.

[9] Much of this section is drawn from Kumbhakar and Lovell (2000).

we truncated the yield variable by assigning all observations below the 5th percentile with the 5th percentile value and similarly constrained those above the 95th percentile to the 95th percentile value.[10] This resulted in a sample mean yield of 2.04 kg of cherry per coffee tree. Average acreage under coffee is calculated at 0.56 while average household total landholdings are several times larger at 1.84 acres. Household and hired labour use, specifically for coffee production, averages 44.8 man days and 29.4 man days per season, respectively. We include the age of the household's coffee trees as a regressor to capture the physiological relationship between tree productivity and its age. A relatively high average age of 31 years indicates a long-standing tradition of coffee production in the area.

Preharvest damage refers to the estimated percentage of total output that was lost before harvest due to natural causes. We recognize the potential endogeneity of this variable as it is likely to be correlated with plot-level managerial ability that also plays a role in determining technical efficiency. However, preharvest damage also contains both exogenous components such as rainfall and pest infestation as well as quasi-fixed components such as plot slope (a key factor, as Murang'a is Kenya's hilliest district). Moreover, not accounting for this variable may significantly overestimate the technical inefficiency of farms struck by exogenous shocks directly impacting on yield. By potentially understating technical inefficiency, we strengthen our final results.

Due to large variance in fertilizer application rates, a significant number of zeros and a range of possible fertilizers to choose from, we constructed a single dummy variable for the use of both inorganic and organic fertilizers. For inorganic fertilizers, we summed up the per kilogram use of DAP, CAN, NPK and MAP, and created a dummy variable equal to 1 if use was above the fifth percentile of all non-zero observations and equal to zero otherwise. In a similar manner, we constructed the dummy for organic fertilizer from use of farmyard manure and compost. With this formulation, 23% and 45% of our sample utilized inorganic and organic fertilizers, respectively. Agricultural productivity depends heavily on agroecological and biophysical conditions that are largely exogenous (Sherlund *et al.*, 2002). In an attempt to capture the effects of agroecological variation on technical efficiency, we included farm-level altitude in the estimation. However, after several different specifications all yielded highly imprecise estimates that often increased the standard errors on the estimates of other model parameters, we opted to leave it out. One possible explanation is that while altitude gradually but steadily increases from the eastern border of Murang'a to its western border at the foothills if the Aberdare Mountain Ranges, its very hilly terrain throughout results in rapid fluctuations of altitude across short distances, which could confound the mean effects.

Assuming that households in close proximity to each other are likely to experience similar agroecological environments, we attempt to overcome

[10] We performed similar transformations for both labour variables and the variable for tree age, which all contained extreme outliers. Without limiting the distribution in this manner, results were less precise and rendered insignificant parameter estimates for several variables that are otherwise significant.

this problem by specifying the inefficiency disturbance to be heteroskadastic across cooperatives. This allows the model to capture cooperative-specific differences in mean technical efficiency. Apart from capturing productivity differences related to environmental conditions, heteroskadasticity across cooperatives will also control for inefficiency effects that arise from poor management at the cooperative level. Among other things, this can include poor timing for the provision of time-sensitive inputs such as fertilizers, credit needed to purchase various inputs (Hanchate, 1996) and political wrangling among the leadership or insecurity that results in effort-reducing distraction among workers (Barrett *et al.*, 2006).

We present the results of our frontier estimation in Table 3.5. Results indicate that the total acreage of land available to the grower, the acreage under coffee, the amount of hired labour used in coffee production, the age of the coffee trees, the use of inorganic fertilizers and the estimated preharvest loss were all significantly related to the observed cherry coffee yield. Given the log-quadratic specification, interpreting the results is not so straightforward. To give a better sense of the estimated relationships, we compute farm-specific elasticities based on actual input levels and frontier parameter estimates. Figure 3.4 presents the kernel regression of the elasticity on variables of interest.

The plot size elasticity of yield is negative and decreases across the whole range of plot size. Essentially, this means that the optimal strategy is not to devote any land to coffee production. Having controlled for total landholdings, increasing plot size under coffee implies a larger percentage of total land devoted to coffee. This provides some evidence (albeit inconclusive) of allocative

**Table 3.5.** Stochastic production frontier estimates.

| Parameter | Coefficient | Standard error | Parameter | Coefficient | Standard error |
|---|---|---|---|---|---|
| Constant | 0.656 | 0.830 | $\ln(\sigma_v^2)$ | −1.853*** | 0.236 |
| Plot size | −0.256** | 0.103 | $\ln(\sigma_u^2)$ | | |
| Plot size$^2$ | −0.043*** | 0.010 | Kamacharia | 1.996*** | 0.758 |
| Land size | −0.138* | 0.079 | Gaturi | 2.597*** | 0.779 |
| Land size$^2$ | −0.050 | 0.039 | Weithaga | 0.025 | 0.805 |
| Household labour | −0.003 | 0.024 | Kanyenyaini | 2.023** | 0.990 |
| Household labour$^2$ | 0.008 | 0.007 | Kahuhia | 2.882*** | 0.801 |
| Hired labour | 0.112*** | 0.020 | Iyego | 2.067*** | 0.803 |
| Hired labour$^2$ | 0.028*** | 0.005 | Kiru | 3.199*** | 0.799 |
| Tree age | 0.638** | 0.263 | Kiriti | 2.870*** | 0.802 |
| Tree age$^2$ | −0.109** | 0.045 | Constant | −1.998** | 0.795 |
| Harvest loss | −0.135* | 0.071 | | | |
| Harvest loss$^2$ | −0.047** | 0.024 | | | |
| Inorganic fertilizers | 0.658*** | 0.112 | | | |
| Organic fertilizers | 0.058 | 0.107 | | | |
| Log pseudo-likelihood | −28011.05 | | No. of observations | 207 | |
| Wald chi$^2$(14) | 210.57 | | Prob > chi$^2$ | 0.00 | |

***Significant at 99% level; **significant at 95% level; *significant at 90% level.

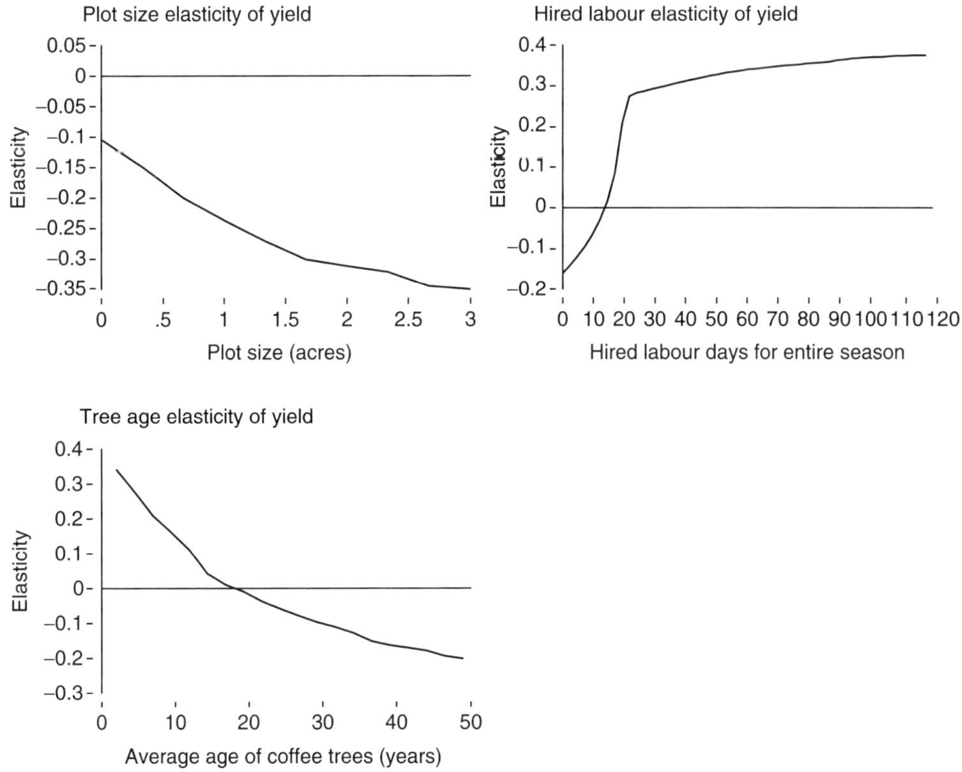


**Fig. 3.4.** Kernel regression of estimated yield elasticities with respect to variables of interest.

inefficiencies, suggesting that available alternatives for land use are likely to be more productive. A majority of farmers, disenchanted by the dismal returns to coffee, indicate that they have recently often neglected their coffee, tending to it minimally or intensively intercropping with beans, napier grass, maize, etc. Since both the required effort and the opportunity costs of land increase with plot size, the larger the plot, the greater the incentives to intercrop and the tendency to reduce husbandry towards a low-return crop. Such a dynamic could also explain the negative plot size elasticity of yield.

While yield response to inputs of hired labour does not follow the typical inverse-U trajectory, it seems to fit the particular context quite well. Coffee production requires a minimum threshold of labour input for any positive level of output. Pruning, weeding, collecting and transporting raw cherry to pulping stations are just some of the necessary tasks. Furthermore, informal labour contracts of less than a day are very rare in Murang'a. Consequently, low inputs of hired labour can imply a high degree of underemployment as a labourer may be needed for several days throughout the season but be largely underemployed. None the less, as marginal labour inputs above the requisite minimum would only occur if current labour time is fully employed, labour elasticity of yield is positive at higher inputs of labour.

The estimated elasticity of yield with respect to tree age is quite intuitive. As coffee trees age to maturity, here estimated at around 18 years, they increase in productivity. However, once they reach maturity their productive capacity peaks and steadily declines thereafter. While this inverse-U age–productivity relationship is expected of the physiological process, it is interesting to note that despite this fact, the uprooting of coffee trees was outlawed until 2002. Furthermore, there is no mechanism within cooperatives that attempts to correct for this by initiating a scheme to facilitate smallholder farmers' replacement of old trees with new ones. As such, the average age of the coffee trees in our sample is over 30, more than 10 years past their prime. The use of inorganic fertilizers, which has dramatically decreased in the recent past as most cooperatives are too heavily indebted to provide such inputs to their members, is strongly and significantly related to yields.

Our principal objective in estimating the production frontier was to obtain the estimates of grower technical efficiency within our sample. We included cooperative-level dummies to control for heteroskadasticity in the inefficiency error term. Estimates for eight of the nine cooperative dummies are statistically significant, indicating that systematic differences in farm-level technical efficiency exist across cooperatives. As presented in Table 3.5, the estimates are parameterized as the log variances of the error components. From these results, we can extract the farm-specific estimates of technical efficiency. These estimates will later be used as the dependent variable for the test we run to look at the effect of corruption at the cooperative level on farm-specific efficiency. Table 3.6 provides some descriptive statistics on the estimates of technical efficiency.

What immediately stands out from these statistics is the large variation in technical efficiencies both within and between cooperatives. Gaturi, whose mean efficiency is slightly less than the full sample average, has farmers posting technical efficiencies as low as 4% and as high as 86% of the optimal. Kangunu and Kanyenyaini are the two most efficient cooperatives with an average efficiency of about 75% of the optimal and very low standard deviations. Kiru, the least efficient, produces at an average rate of just 26% of the optimal. Across the whole sample, farmers post an average efficiency rate of 50%.

**Table 3.6.** Descriptive statistics of technical efficiency estimates.

| | Mean | Standard deviation | Minimum | Maximum |
|---|---|---|---|---|
| Kamacharia | 0.50 | 0.19 | 0.18 | 0.78 |
| Gaturi | 0.46 | 0.25 | 0.04 | 0.86 |
| Weithega | 0.75 | 0.06 | 0.66 | 0.87 |
| Kanyenyaini | 0.59 | 0.21 | 0.07 | 0.84 |
| Kahuhia | 0.39 | 0.24 | 0.04 | 0.80 |
| Iyego | 0.54 | 0.21 | 0.10 | 0.85 |
| Kiru | 0.26 | 0.17 | 0.06 | 0.57 |
| Kangunu | 0.76 | 0.08 | 0.64 | 0.88 |
| Kiriti | 0.42 | 0.28 | 0.05 | 0.88 |
| Total | 0.50 | 0.25 | 0.04 | 0.88 |

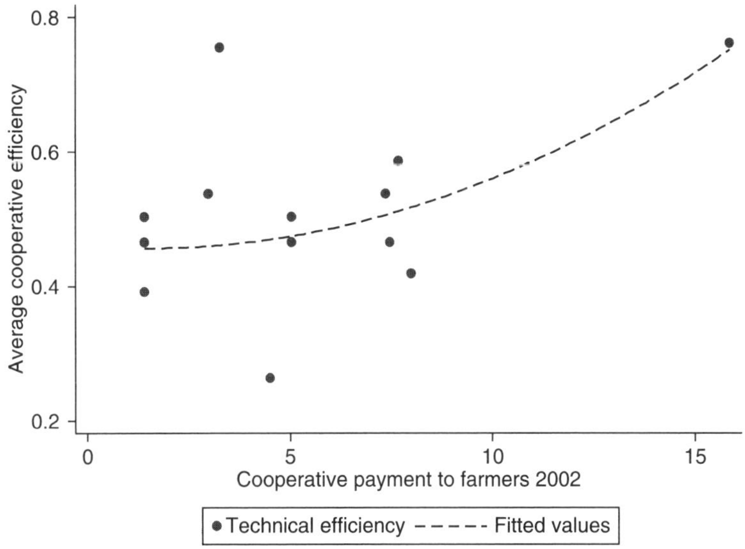


**Fig. 3.5.** The relationship between cooperative payments to farmers and their technical efficiency.

Although, for reasons previously explained, we used variations in technical efficiency to confirm the existence and consequence of rent-seeking behaviour on cooperative performance, our analytical model made no explicit predictions of the relationship between corruption and technical efficiency. Instead, it showed that where corruption was present, payments made to farmers would be relatively lower. We further justify the use of technical efficiency as an indicator of rent-seeking activities by showing, in Fig. 3.5, the relationship between payments and mean efficiency at the cooperative level. As expected, lower payments are associated with lower levels of efficiency. The results suggest that technical efficiency can be used as a proxy for lower payments, whereby low levels of technical efficiency imply a greater likelihood of corruption amongst the cooperative leadership.

## 4.4 Creating proxies associated with rent-seeking behaviour

The next step is to create proxies as indicators for the relative level of corruption and management incompetence between the cooperatives. The idea is to identify and differentiate cooperatives by the likelihood that they are run by rent-seeking or inept board members. This is not a straightforward task as there is no direct measure of corruption; nor would the cooperative management offer any information that may implicate them. Moreover, our sample includes only 9 cooperatives with a total of 13 factories limiting any regression analysis we can use to generate the requisite proxies.

As such, we take a different approach, seeking to identify corruption within cooperatives indirectly by studying the outcomes and perceptions that are commonly associated with corruption or mismanagement. It is important to note that with the data we have available, we are not able to differentiate between corruption and management incompetence. Nevertheless, as the negative effect that either corruption or mismanagement has on member welfare is identical, we feel that making such a distinction is not crucial to our analysis or its implications. Henceforth, we use corruption and mismanagement interchangeably.

Factor analysis, which is concerned with uncovering the latent structure of a set of variables, is well suited for our purposes. Essentially, factor analysis uses the covariance matrix generated from a set of variables to find a smaller number of common factors that linearly reconstruct the original variables. These common factors, depending on the weights they assign to the different variables (known as 'loadings'), can then be interpreted as proxies for the common structure that they represent.

We use factor analysis to reduce a set of variables into common factors which correspond to various aspects of cooperative organization and practice that are likely to affect their productivity and are plausibly related to the degree of corruption plaguing the cooperatives. We conduct two separate tests, extracting two underlying factors from each. The first includes only factory-level variables and aims at generating common factors that relate to the structure and performance of the cooperatives. The second test includes farmer-level variables that capture subjective perceptions of cooperative effectiveness, and farmers' confidence with cooperative policy makers and government officials.

Table 3.7 defines the variables used in the first test and provides some basic statistics. Payments to farmers, which are unique at the factory level, averaged a mere KSh 5.52 for the 2002/03 season.[11] Cooperatives averaged 3412 active members and owned just over five factories on average. To capture the effect of externalities associated with variations in intracooperative factory payments, we include their coefficient of variation, which has a mean value across cooperatives of 0.31. The net value of coffee sales, which is the factory-specific price received at the Nairobi Coffee Auctions (where the vast majority of coffee is marketed) less marketing and milling costs, proxies for

**Table 3.7.** Variables used to generate size and performance factors for cooperatives.

| Variable | Definition | Mean | Standard deviation |
|---|---|---|---|
| Payment | Total factory level payment (KSh/kg of output) made to members for 2002/03 season | 5.52 | 3.98 |
| Members | Number of active members per cooperative | 3413 | 1738 |
| Factories | Number of factories operated by cooperative | 5.31 | 3.12 |
| Payvariance | Coefficient of variation of intracooperative pay | 0.31 | 0.4 |
| Quality00 | Net value of coffee sales 2000 (KSh/kg) | 81.69 | 28.85 |
| Quality99 | Net value of coffee sales 1999 (KSh/kg) | 76.68 | 8.03 |
| Coopyield | Average cooperative yield for 2002/03 season (kg/tree) | 1.84 | 0.67 |

[11] Recall that all prices and values are deflated to 1998 prices.

**Table 3.8.** Results of cooperative size and performance factor analysis.

| Factor | Eigenvalue | Proportion of variance explained |
|---|---|---|
| Size | 2.98 | 0.54 |
| Performance | 1.70 | 0.31 |

ROTATED FACTOR LOADINGS

| Variable | Size | Performance | Uniqueness |
|---|---|---|---|
| Members | 0.97 | –0.21 | 0.02 |
| Factories | 0.96 | –0.19 | 0.04 |
| Payvariance | 0.90 | 0.16 | 0.15 |
| Payment | –0.22 | 0.87 | 0.19 |
| Quality00 | –0.04 | 0.58 | 0.66 |
| Quality99 | 0.32 | 0.29 | 0.81 |
| Coopyield | –0.06 | 0.75 | 0.44 |

the quality of coffee produced. We also include the average cooperative yield, which for the 2002/03 season was a paltry 1.84 kg of cherry per tree.[12]

In Table 3.8 we present the factor analysis results. Of seven possible factors, we retain only two for extraction. There is no strictly defined procedure for determining the optimal number of factors to retain. The decision often depends on a combination of a predetermined hypothesis and a sensible cut-off criterion; either a threshold eigenvalue above which factors are retained (commonly between 0 and 1) or a degree of variation that the first *n* factors must explain (commonly 90%).

The first factor, which we call *size*, describes the cooperative size. It loads heavily on members, factories and variance of pay within the cooperative, which are all correlated with increasing size. The low uniqueness posted by each of these variables indicates that the underlying factor of size is well defined by these variables. Uniqueness is defined as the fraction of variance for the variable that is *not* explained by the factors. The second factor, which we call *performance*, loads primarily on variables associated with the performance or productivity of the cooperative as given by the payments its members receive, the volume of output they produce and the quality of their coffee. As one would expect, and as indicated by the large positive loadings of these variables on the *performance* factor, performance improves as payments, production and quality of coffee increase.

Beyond the structural features of a cooperative that may determine the ease with which corruption takes root, or its relative performance that may proxy for the extent of mismanagement or rent-seeking activity, members' beliefs regarding the effectiveness of cooperative management could also reveal

[12] As we will subsequently be using the factors generated from this analysis as explanatory variables in a regression with the extracted technical efficiency estimates as the independent variables, we exclude own yield from the average cooperative yield generated for each household. This avoids the problem of spurious correlation that could arise since the technical efficiency estimates are also based on yields.

**Table 3.9.** Variables used to generate members' perception and confidence factors.

| Variable | Definition | Response structure |
|---|---|---|
| Goodrelations | Members of the coop have generally good relationships with each other | 1 |
| Caninfluence | Membership can influence decision-making process | 1 |
| Profitdistribute | Membership understands how management distributes cooperative profits | 1 |
| Effectivemanage | Management is effective in running the cooperative | 1 |
| Coopcompare | Your cooperative is managed better than other coffee cooperatives in the region | 1 |
| Insecurity | This village/neighbourhood has a problem with insecurity and violence | 1 |
| Localgovt | Local government officials can be trusted | 1 |
| Centralgovt | Central government officials can be trusted | 1 |
| Agofficer | District agricultural officers do their best to improve the welfare of farmers | 1 |
| Coopofficer | District cooperative officers do their best to improve the welfare of farmers | 1 |
| Creditaccess | Do you have access to money-lending facilities | 2 |
| Empowerment | Are you able to make important decisions that could change the course of your life | 3 |

**Response structure**

| | |
|---|---|
| 1 | 1 = Strongly agree; 2 = Agree; 3 = Neither; 4 = Disagree; 5 = Strongly disagree |
| 2 | 1 = Yes; 2 = No |
| 3 | 1 = Totally unable; 2 = Largely unable; 3 = Neither; 4 = Largely able; 5 = Totally able |

key information. To investigate this possibility, we ran a second factor analysis on the variables defined in Table 3.9. In order to facilitate interpretation of the subsequent factor loadings, we include the variables' response structure.

Note that all the variables used for the second factor analysis are ordinal. While ordinality presents a theoretical problem for factor analysis, which is developed for continuous variables, methods created to account for ordinality have thus far proved to be largely computationally infeasible (Joreskog and Moustaki, 2001). The limited set of methods that can incorporate ordinality into factor analysis, known generally as the underlying bivariate normal (UBN) approach, has proven to yield similar or only slightly better results compared to 'standard' ordinal coding. Kolenikov and Angeles (2004) argue that the modest gains from using computationally intensive methods to control for ordinality are not sufficient to recommend its use. Where the variables are mostly dummies, however, they show that not controlling for discreteness leads to significantly inferior results. For completeness, we tried to correct for ordinality by using the *polychoricpca* function in STATA. The two factors thus generated were difficult to interpret and furthermore, when they were introduced as variables in the technical efficiency

**Table 3.10.** Factor analsysis of members' perception and confidence.

| Factor | Eigenvalue | Proportion of variance explained |
|---|---|---|
| Dissatisfaction | 1.48 | 0.60 |
| Pessimism | 0.82 | 0.33 |

ROTATED FACTOR LOADINGS

| Variable | Dissatisfaction | Pessimism | Uniqueness |
|---|---|---|---|
| Goodrelations | 0.23 | 0.04 | 0.95 |
| Caninfluence | 0.24 | 0.05 | 0.94 |
| Profitdistribute | 0.37 | 0.03 | 0.87 |
| Effectivemanage | 0.64 | 0.03 | 0.58 |
| Coopcompare | 0.61 | 0.00 | 0.62 |
| Creditaccess | 0.19 | –0.06 | 0.96 |
| Insecurity | –0.22 | –0.06 | 0.95 |
| Coopofficer | 0.40 | 0.32 | 0.73 |
| Agofficer | 0.24 | 0.40 | 0.78 |
| Localgovt | 0.05 | 0.56 | 0.68 |
| Centralgovt | –0.05 | 0.55 | 0.69 |
| Empowerment | –0.06 | 0. 23 | 0.94 |

estimate (substituting for the two factors, *perception* and *confidence*, generated without correcting for ordinality), the model's R-squared dropped and neither factor was significant. As such, we opted against controls for ordinality.

In Table 3.10, we present the results of the second factor analysis. Again, we retain only the first two factors, which together explain over 90% of the variance among the variables. The first factor, *dissatisfaction*, loads heavily on variables that encompass subjective beliefs of how well the cooperative is managed. These include member ability to influence policy, member understanding of how cooperative resources are used, member ability to access credit and even member beliefs of the dedication of government cooperative officials to their job. While the uniqueness levels of most of these variables are high, *effectivemanage* and *coopcompare*, which are closely related to our interpretation of the underlying factor, are associated with acceptable levels of uniqueness. As loaded, *dissatisfaction* increases as farmers are more likely to rate their cooperative as poorly managed, lacking in the provision of services, associated with violence and disengaged from the membership. As such, we would expect *dissatisfaction* to be negatively related to farmer technical efficiency.

The second factor seems to represent a measure of pessimism in the regulatory environment or a lack of faith in the commitment of policy makers to improve cooperative performance. We call this factor *pessimism*. It is instructive to note that in December 2002, about a year before the fieldwork for this research took place, Kenya held a landmark national election that resulted in the first electoral transfer of power in the country's history. The sense of unbounded optimism and hope was palpable and permeated the whole country

(Wolf *et al*., 2004).[13] In conversations with our respondents, it was clear that some of this confidence remained. Several respondents attributed their expectation of improved performance in the coffee smallholder sub-sector to their confidence in the new government and its commitment to economic growth. Such sentiment is likely to generate incentives for heightened productivity. Thus we expect a negative correlation between *pessimism* and technical efficiency. Note that the variable *empowerment* loads inconsistently on *pessimism*. Unlike the other variables, higher values of *empowerment* are associated with greater self-assurance and should therefore load negatively on *pessimism*. Nevertheless, with a uniqueness level of 0.94, the variable *empowerment* provides little information to the factor *pessimism*.

## 4.5 Determinants of farm-specific technical inefficiency

The third empirical test regresses the farm-specific estimates of technical inefficiency, on a set of likely covariates, including the factors associated with various aspects of rent-seeking, in an attempt to determine the correlates of inefficiency. Table 3.11 presents some descriptive statistics of the variables used in the regression.

We control for the traditional household demographic variables in addition to experience (the number of years the household has been growing coffee), the receipt of extension services, as well as the ratio of advance payment to total payment received by members. About 82% of household heads are male and the mean age for a household head is approximately 58. Growing coffee is very much an established tradition in the area with the average household engaged in coffee production for over 25 years. The percentage of household heads that

**Table 3.11.** Descriptive statistics for correlates of inefficiency regression.

| Variable | Mean | Standard deviation | Minimum | Maximum |
|---|---|---|---|---|
| Gender (male = 1) | 0.825 | 0.381 | 0 | 1 |
| Age (years) | 57.753 | 13.758 | 27 | 96 |
| Household size | 5.096 | 2.244 | 1 | 15 |
| Primary education | 0.332 | 0.472 | 0 | 1 |
| Secondary education | 0.328 | 0.470 | 0 | 1 |
| Postsecondary | 0.057 | 0.232 | 0 | 1 |
| Experience (years) | 26.017 | 12.762 | 1 | 65 |
| Extension (received in last 2 years = 1) | 0.320 | 0.467 | 0 | 1 |
| Advance payment ratio 2002 | 0.519 | 0.308 | 0.133 | 1.00 |
| Size | 0.00 | 0.99 | −1.12 | 2.32 |
| Performance | 0.00 | 0.92 | −1.46 | 2.09 |
| Dissatisfaction | 0.00 | 0.81 | −2.51 | 1.70 |
| Pessimism | 0.00 | 0.75 | −1.93 | 1.93 |

[13] Gallup International, in their annual end-of-year survey, ranked Kenya as the most optimistic country in December 2002.

have completed either primary or secondary school stands roughly at 33%, but less than 6% of household heads have any postsecondary schooling. Only 32% of the households had been visited by an extension agent at least once in the last 2 years. For the 2002/03 season, the average ratio of advance to total payments stood just above 50%. The advance ratio is defined as the fraction of total payment that is received as an advance at the start of the season.[14] We include this variable as a proxy for liquidity.

Various hypotheses exist to explain the existence and direction of the relationship between the chosen demographic covariates and technical efficiency: gender differentiation in farming activities, the inertia of older households to novel and superior farming practices, the benefits to specialization of roles accorded to larger households, etc. However, as these relationships are not the focus of this study, we remain largely agnostic of their role. As extension services are an important service that cooperatives specifically provide their members to improve their productivity, we expect the receipt of such services to be positively associated with technical efficiency. Furthermore, as advance payments would allow farmers to invest in quality inputs in a timely manner, we hypothesize a positive relationship with technical efficiency.

To test for the relationship between cooperative corruption and farmer technical efficiency, we include the four factors associated with different aspects of cooperative organization and performance that can be linked to the likelihood and extent of rent-seeking activity. The actual values of these variables do not mean much and are here simply normalized to mean zero. However, because the variables are cardinally ranked, the position of a given observation relative to the variable's entire range is important. Given that the variables loading heavily on both the *size* and *performance* factors are all positive, and are all positively related to the underlying notion of cooperative size and performance (e.g. more members and more factories likely signify a larger cooperative as larger payments to members and greater production indicate a better-performing cooperative), higher values of these factors suggest bigger and more efficient cooperatives. *Dissatisfaction* and *pessimism*, on the other hand, are both negative, increasing as members perceive greater management incompetence or have less confidence in policy makers.

None of the demographic variables, including experience in coffee growing, proves to be significantly related to degree of efficiency. A possible explanation is that some of these variables are related to the use and availability of inputs whose variation is already captured in the estimates of technical efficiency (Table 3.12). Household size, for example, is probably associated with the use of household labour and the demand for hired labour. Experience, on the other hand, is likely to be correlated with tree age.

The receipt of extension services, which we expected to be associated with greater levels of efficiency, does appear to have a statistically significant effect. Given that the mean value of the technical efficiency estimates was a mere 0.50, the high value of the extension dummy's estimated parameter

[14] Recall that cooperatives pay their members in two installments: a CAPS at the beginning of the season and a final payment at the end of the season.

**Table 3.12.** Sources of inefficiency estimates.

| Variable | Coefficient | Standard error |
|---|---|---|
| Constant | 0.239 | 0.285 |
| Gender | –0.024 | 0.049 |
| Age | 0.003 | 0.010 |
| Age$^2$ | –1.1E–05 | 8.3E–05 |
| Household size | 0.001 | 0.008 |
| Primary education | 0.017 | 0.050 |
| Secondary education | –0.021 | 0.056 |
| Postsecondary | 0.001 | 0.082 |
| Experience | –0.001 | 0.006 |
| Experience$^2$ | –6.0E–06 | 1.0E–04 |
| Extension | 0.059* | 0.035 |
| Advance payment ratio 2002 | 0.356*** | 0.103 |
| Size | –0.031* | 0.017 |
| Performance | 0.179*** | 0.034 |
| Dissatisfaction | 0.037* | 0.022 |
| Pessimism | –0.040* | 0.023 |
| $R^2$ | 0.2069 | |
| Number of observations | 197 | |

***Significant at 99% level; *significant at 90% level.

emphasizes the importance of such services. Receiving extension services at least once in the last 2 years increases technical efficiency by slightly more than 10%. That a mere 32% of households have had extension visits in the last 2 years is further evidence of the deterioration of coffee cooperatives.

The fraction of the total payment given as an advance at the beginning of the season, a key policy variable, is significantly and positively related to efficiency. This points to the crucial importance of providing smallholder farmers, who are often cash-constrained and have limited access to credit, with some form of advance payment on their output in order to facilitate the timely purchase of critical inputs such as inorganic fertilizers and pesticides.

The four cooperative-specific factors are all significantly associated with farm-level technical efficiency. The *size* factor is negatively associated with technical efficiency. On average, the larger a cooperative is, the less technically efficient are its members. This result is consistent with election capture given traditional voting methods and no controls against bribery. As we showed, the likelihood of election capture increases with increasing membership, a variable clearly related to a cooperative's size. As such, this result could be interpreted as revealing a significant association between the probability that a cooperative has been captured by rent-seeking officials and technical efficiency. This finding can also be interpreted as suggesting decreasing returns to scale in cooperative production, a result that contradicts the unsupported claims of many Kenyan policy makers who champion mergers in cooperatives on the basis of increasing returns to scale arguments.

As the issue of scale is an important policy question, we supplement these results with further evidence in support of decreasing returns to scale.

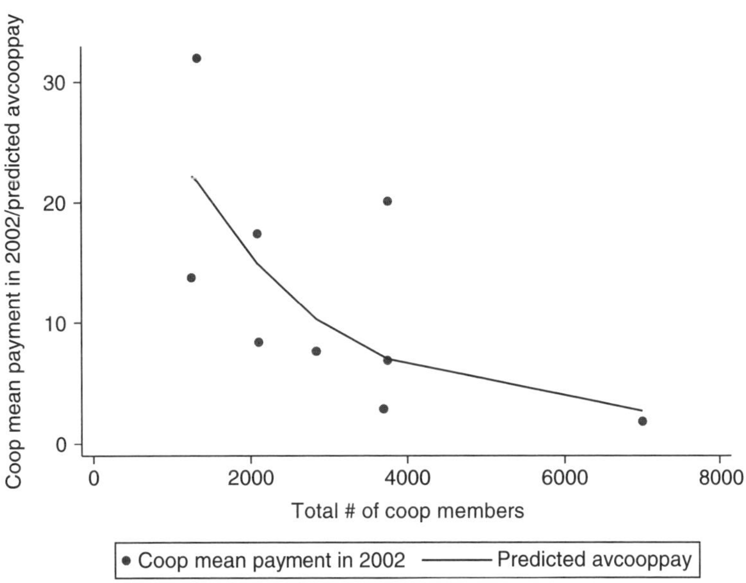


**Fig. 3.6.** Mean cooperative payments by total members per cooperative.

Figure 3.6 presents a scatter plot of the relationship between mean cooperative level payments given to farmers with the total number of members in the cooperative. As the non-parametric prediction shows, the clear negative relation supports diminishing returns to scale, where returns are proxied by farmer earnings.

However, as Fig. 3.7 suggests, size is not necessarily detrimental to efficiency. That mean cooperative payments are actually increasing in the average number of members per factory implies increasing factory-level economies of scale. As elections occur at the cooperative level, these results together suggest that while economies of scale in the production and marketing of coffee are manifest in increasing returns at the factory level, the increasing likelihood of election capture is sufficient to negate the benefits of scale at the cooperative level.

Returning to the model results, the negative and significant result on *pessimism* lends further credence to our claims. Lack of confidence in policy makers' resolve to improve the rules and regulations that underlie the smallholder coffee sector suggests a current institutional arrangement that does not provide growers with incentives that aptly reward productive behaviour.

The only unexpected result regards the positive and significant relationship between *dissatisfaction* and efficiency. Because *dissatisfaction* explains the least variation among the variables it loads heavily on, it could be that our interpretation of the latent variable captured in the variable *dissatisfaction* is somewhat imprecise. An alternative explanation is that having controlled for cooperative performance, *dissatisfaction* may be picking up farmer-specific expectations of how cooperatives could be managed, which in turn is associated with a farmer's understanding of the disparity between the status quo and what is possible under optimal management. Such farmers, who are

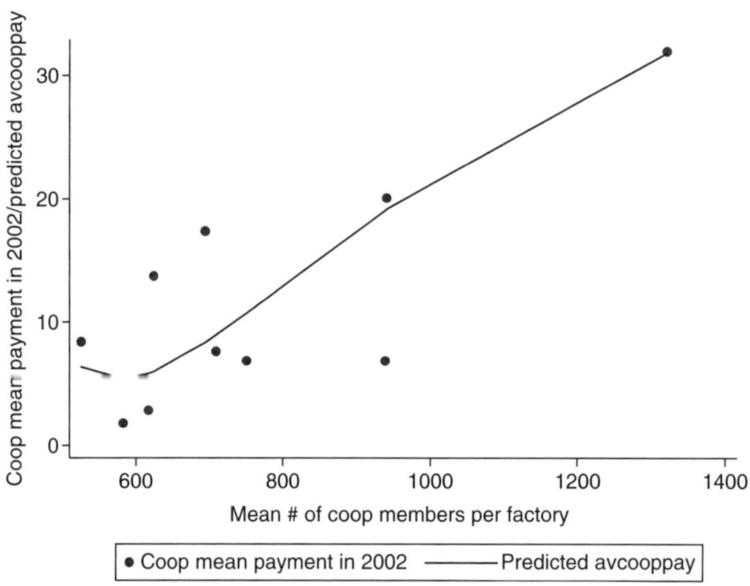


**Fig. 3.7.** Mean cooperative payments by total members per cooperative.

likely to be the most enterprising, would express the most discontent with management while still applying effort into their own production. High values of *performance*, associated as it is with higher payments to farmers and increases in the quality and quantity of output, suggests a cooperative leadership that seeks to maximize member welfare and provide the right incentives for increased productivity. Low values of *performance*, on the other hand, are more likely to be associated with the fleecing of cooperative revenues by the leadership and the neglect of administrative and management duties. Indeed, our analytical model showed that cooperatives captured by a rent-seeking leadership would be characterized by lower payments to farmers and lower farmer output. Cooperatives with low values of *performance* are thus more likely to be headed by corrupt or inept individuals. As such, the sizable and strongly significant positive relationship between *performance* and farmer efficiency also supports the hypothesis that rent-seeking at the cooperative level impacts negatively on farm-level efficiency.

## 5 Discussion and Conclusion

For almost a decade now, the smallholder coffee sub-sector in Kenya has witnessed a significant deterioration in several key indicators. Both yields and total output have declined, payments to farmers have plummeted and the relative price of Kenyan coffee in the world market has declined. In this chapter we investigated the relationship between this decline and changes in the laws governing cooperatives that effectively shifted the managerial and administrative responsibility of running a cooperative onto its members. We argued that where managers are chosen from the membership, and where no regulatory oversight exists to curb election fraud or the embezzlement of collective profits, leadership positions are prone to capture by corrupt individuals who have a greater incentive to win elections than benevolent candidates have. We supported our argument by highlighting deficiencies in the voting procedures used to elect cooperative leaders, and in the law that requires members to market their coffee solely through their cooperatives. Together, these two institutional features promote election capture by facilitating vote-buying, and increasing the incentives for rent-seeking by forcing members to acquiesce to exploitation.

Having set the stage, the empirical part of this study seeks to identify and rank cooperatives by the likelihood that they have been captured by a corrupt and rent-seeking leadership. Using factor analysis to create cooperative-level proxies for corruption or ineptitude, we estimate the effect that these factors have on farm-level technical efficiency and find that the same measures of cooperative structure, performance and member perception associated with increased rent-seeking are also significantly related to reduced technical efficiency. In other words, the evidence suggests that corruption and incompetence among the governing board breeds disillusionment among members that manifests in declining technical efficiency.

Other key results confirm the significant and positive impact that fertilizers, pesticides and extension services have on coffee yields. Given the extremely low fertilizer and pesticide application rates evident in the sub-sector, and a weak extension system, mechanisms designed to boost the use of these inputs could have a large, favourable impact on yields.

These results suggest several policy implications. First, efforts need to be taken to ensure that politically savvy and self-interested individuals do not continue to manipulate elections. A feasible step towards this end would be to require that elections are carried out by secret ballot in the presence of objective cooperative election supervisors. Second, in order to pressure cooperatives to operate efficiently and to provide growers with alternatives, growers should be allowed to sell their output to the highest bidder. The current circumstance of coffee cooperatives presents a golden opportunity to make this shift into competitiveness, as the main justification for involuntary membership no longer applies. When cooperatives used to provide their growers with credit, fertilizers, pesticides, extension advice, etc., requiring growers to sell their output to their respective cooperatives was a credible means to ensure repayment for services rendered and inputs provided. As cooperatives today rarely provide these inputs or services, this argument is moot. Increased competitiveness, however, would require a legal system that can formally enforce contracts.

The main goal of organizing smallholder coffee growers into fully autonomous cooperatives was to create the economic conditions that would encourage their productive and marketing capacities. Divestment of government responsibility in cooperative management was meant to empower members, maximize the return to farmers' efforts, increase their output and ultimately improve their welfare. Unfortunately, scant attention was given to the incompatibility of the socio-economic and institutional apparatus onto which the programme of market decentralization of cooperatives was to be grafted. The lesson is clear: giving full ownership of producer organizations to members without complementary and enabling regulations, or when the potential for information manipulation exists, only creates the illusion of empowerment and, as we have seen among Kenya's coffee cooperatives, can lead to a deterioration of members' welfare.

## Acknowledgements

I thank Miriam Chirwa, Ravi Kanbur, Andrew Karanja, Raphael Gitau, Mwangi Githinji, Ester Mwiruri, James Nyoro, John Omiti, Paulo Santos, Raphael Thuo, Christoph Vanborg, seminar participants at Cornell University and the World Bank Nairobi, as well as participants at the 2005 USAID SAGA Conference in Nairobi for all their helpful comments, insights and assistance. A special thanks to Chris Barrett for extensive assistance throughout the project. Many thanks to my field enumerators for their diligent effort, and a special thanks to all the survey respondents who freely gave their time to offer valuable information, and to the Institute of Policy Research and Analysis and Tegemeo Institute for Agricultural Research and Policy for providing institutional and logistical support for data collection in Kenya. This research was partly funded through USAID's Strategies and Analysis for Growth and Access (SAGA) programme, the USAID Pastoral Risk Management (PARIMA) programme under grant PCE-G-98-00036-00 and Rockefeller Kenya. The views expressed are solely the author's and do not represent any official agency. Any remaining errors are mine alone.

## References

Aigner, D.J., Lovell, C. and Schmidt, P. (1997) Formulation and estimation of stochastic frontier production function model. *Journal of Econometrics* 6, 21–37.

Barrett, C.B., Sherlund, S. and Akinwumi, A. (2006) Macroeconomic shocks, human capital and productive efficiency: evidence from West African farmers. *Journal of African Economies* 15, 343–372.

Ephanto, R.K. (2003) The impact of Kenya's foreign exchange rate policy on agricultural export incentives. In Conference Proceedings on Agricultural Exports and Market Development, PAM, Nairobi.

Gugerty, M. and Kremer, M. (2004) The Rockefeller effect. Massachusetts Institute of Technology Poverty Lab Working Paper.

Hanchate, A.D. (1996) Semiparametric estimation of yield loss due to mistiming of crop operations. *Economic Letters* 51, 253–257.

Howes, M. (1997) NGOs and the institutional development of membership organizations: a Kenyan case. *The Journal of Development Studies* 33, 820–848.

Joreskog, K. and Moustaki, I. (2001) Factor analysis of ordinal variables: a comparison of three approaches. *Multivariate Behavioral Research* 36, 347–387.

Karanja, A.M. (2002) Liberalization and smallholder agricultural development: a case study of coffee farms in Kenya. PhD thesis, Wageningen University. ISBN 90-5808-603-8.

Karanja, A.M. and Nyoro, J. (2002) Coffee prices and regulation and their impact on livelihoods of rural community in Kenya. Tegemeo Institute Working Paper, Nairobi, Kenya.

Kolenikov, S. and Angeles, G. (2004) The use of discrete data in PCA: theory, simulations, and applications to socioeconomic indices. UNC Chapel Hill, Working Paper 04-85, Chapel Hill, North Carolina.

Kumbhakar, S. and Lovell, C. (2000) *Stochastic Frontier Analysis*. Cambridge University Press, Cambridge.

Meeusen, V. and van den Broeck, J. (1977) Efficiency estimation from Cobb–Douglas production functions with composed error. *International Economic Review* 18, 435–444.

Morduch, J. (1999) The microfinance promise. *Journal of Economic Literature* 37, 1569–1614.

Rahman, A. (1999) Micro-credit initiatives for equitable and sustainable development: who pays? *World Development* 27, 67–82.

Sherlund, S.M., Barrett, C.B. and Adesina, A. (2002) Smallholder technical efficiency controlling for environmental production conditions. *Journal of Development Economics* 69, 85–101.

Stiles, K. (2002) International support for NGOs in Bangladesh: some unintended consequences. *World Development* 30, 835–846.

Wolf, P., Logan, C. and Owiti, J. (2004) A new dawn? Popular optimism in Kenya after the transition. Afrobarometer Working Paper No. 33.

# 4 Decentralization, Community-based Organizations and Access to Agricultural Extension Services in Eastern Kenya

G.L. MUGUNIERI AND J.M. OMITI

## 1 Introduction

### 1.1 Background

The importance of agricultural extension in rural development is widely acknowledged, particularly in developing countries where a majority of the population lives in rural areas and agriculture is the main source of livelihood (Wanga, 1999). Agricultural extension, however, has frequently not generated its expected and necessary impact. Various reasons have been advanced for this failure. These include farmers' ignorance, apathy and lack of local leadership (1950s and 1960s); existence of farm-level constraints such as credit and physical infrastructure (1970s); lack of participatory processes leading to economic decline and social disintegration (1980s); and lack of appropriate technologies, befitting the needs of potential adopters (1990s) (Chambers, 1993). Recently, the failure of extension services has been attributed to poor governance and a lack of institutional innovations to ensure greater efficiency and accountability in the mobilization, organization and control of resources (Amudavi, 2003).

These failures have caused many developing countries to decentralize agricultural extension services, with the expectation that these services would become more accessible and more accountable to farmers, thereby improving the flow of information and stimulating technology uptake, market participation and productivity growth. Decentralization is conceptualized as involving the dynamics of reorganizing government services delivery, by transferring responsibilities to more local (governmental or non-governmental) institutions with the objective of improving relevance and responsiveness to users (Smith, 1997). There are at least three reasons why governments have been decentralizing agricultural services (Smith, 1997):

**1.** The growing desire to roll back the role of the state due to past failures of central governments to handle well the complexities of locality-specific issues;
**2.** The inability of governments to continue to finance a whole range of services;
**3.** The view that democracy is best served through devolved functions with enhanced participation at local level.

Although different forms of decentralization exist, two forms have been in operation in Kenya in recent years. The first is decentralization of government responsibility for extension through structural reform, with the objective of shifting extension to other institutions. This has been aimed at improving institutional responsiveness and accountability (Antholt, 1994). The second is decentralization of management of programmes through farmer participatory involvement in decision making and, ultimately, taking responsibility for extension programmes (Amudavi, 2003).

This study explores whether decentralization of extension services enhances access for poor farmers, especially for demand-driven extension services that respond directly to farmers' needs and encourage active farmer participation. We focus on one region of eastern Kenya within which decentralization has progressed to markedly differing degrees, enabling comparison within a similar jurisdictional and agroecological zone. By studying how decentralization impacts the availability and effectiveness of extension provision, and the relationship between community-based organizations (CBOs) and access to, and quality of, extension services, these research results can inform strategies and policies that lead to the enhanced access and effectiveness of agricultural extension services.

## 1.2 Research problem and justification

A critical look at the process of decentralization of extension services in Kenya reveals marked differences in experience. This process appears to have been relatively successful for major crops where the responsibility of farmer education and dissemination of agricultural information was transferred to various development authorities, boards, cooperatives and factories. Some of these agencies include the Horticultural Crops Development Authority, Kenya Tea Development Authority and cooperative societies of some major crops like coffee, among others. This transfer of responsibilities has contributed to farmer participation in the running and management of extension programmes and in meeting the cost of the extension services, implying that these services have been decentralized and privatized.

By contrast, the low-value or traditional food crops sector has benefited much less, if at all, from decentralization. These crops are important for food security. Livestock production within the pastoral and agropastoral areas is another relatively neglected sector. There is need to harness the potential of this sub-sector through proper management and education. This should be geared not only towards increased production but also towards ensuring that livestock products from these areas are disease-free and meet the *Office Internationale des Epizooties* standards for entry into the world market. Last but not least, there is the smallholder dairy sub-sector that contributes 60–80% of Kenya's milk output and owns about 83% of its dairy cattle. Previously reliant on the public sector for extension support, the extension infrastructure these farmers are faced with following the collapse of government support is not well understood.

These three sub-sectors, if supported by effective extension machinery, could provide secure livelihood for many of the nation's farmers and enhance food security. However, with entrenched poverty, it is unclear whether poor farmers can afford privatized extension services. The government, international donors and interested agricultural development agencies thus need to identify workable extension service delivery models that would be promoted for enhanced growth of these sub-sectors.

This study explores whether, and to what extent, decentralization of agricultural extension has enhanced access and participation for farmers within the smallholder, traditional sub-sectors. More specifically, the thrust is to:

**1.** Explore farmers' awareness of different agricultural extension service delivery channels across divisions with different experiences of decentralization;
**2.** Identify and document the quality, cost and accessibility of different agricultural extension delivery channels existing following decentralization;
**3.** Identify the factors associated with access to extension services and the role of CBOs in facilitating such access;
**4.** Draw out policy implications from the findings generated by the study.

## 1.3 Conceptual framework

The reorganization of agricultural extension services in Kenya provides an example of decentralization in a difficult context, partly due to lack of information on its possible diverse impacts on the resource-poor farmers and partly due to lack of a comprehensive institutional framework to guide the process as well as the content. After the structural adjustment programmes of the 1990s, extension provision has evolved to include four broad forms of delivery systems (Anderson and Van Crowder, 2000):

**1.** *Public delivery and public finance*: This comprises traditional government agricultural extension services based on research station–extension agents–farmer linkages. This channel has, however, been constrained by lack of funds and the growing inability of the state extension services to effectively provide for farmers. Extension delivery through the mass media (radio programmes) may also fall under this category.
**2.** *Public delivery and private finance*: This is a form where government staff can be contracted by private agencies.
**3.** *Private delivery and private finance*: This form describes private extension provision with little or no government participation, such as commodity outgrower schemes, or delivery through producer associations. This is predominantly linked to commercialized agriculture and thus does not cater

much to low-income producers and basic staple commodities, though it may benefit the poor as consumers and labourers. Other examples of this delivery system include the outlets selling agricultural inputs, popularly known as agrovets.

**4.** *Private delivery and public finance*: This approach is an essential element of reforming the extension services. It entails outsourcing responsibility for extension delivery to private sector providers, e.g. non-governmental organizations (NGOs) and CBOs. This channel of involving NGOs and CBOs has emerged as an important pathway, with several comparative advantages over the other channels, including grass-roots contacts and use of participatory methods. International donors did not initially recognize and fund NGOs or include them in the research and development process (Hangrave, 1999). However, following the structural adjustment programmes of the 1990s, donors became interested in NGOs since they were private entities. This shift in development thinking strengthened the move towards decentralization and privatization, resulting in more attention being given to NGOs, which now play a major role in the delivery of extension services in Kenya.

These four forms of extension service delivery interact and influence household objectives and output. The government was the original provider but with its failure to reach many farmers effectively, private and bilateral donors came alongside to assist the government. However, farmers' problems with regard to access and participation have persisted. When development thinking turned to the liberalized market as an answer to farmers' problems, private companies and NGOs moved in. More recently, the recognition that farmer participation is a crucial element of an effective extension system has led to the rise of CBOs as a major player. The major players in extension today are the government as the public provider, private companies as private providers, NGOs with donor support working with farmers as well as CBOs as publicly funded, community-based providers, and the media.

One of the major challenges facing extension in Kenya is how to reorient and renew extension with an emphasis on partnership, participation and sharing of knowledge and information while at the same time balancing continuity of extension service provision (Kenya, 2001c). All four forms of extension delivery channels exist today in Kenya, sometimes all in a single geographic area, and interact in a variety of ways with other economic and institutional factors to influence households' decisions, output and welfare as illustrated in Fig. 4.1.

## 2 Methodology

### 2.1 Study area and sampling

This study was undertaken in Makueni district, one of the 12 districts in Kenya's eastern province. Makueni district covers an area of 7440 km$^2$. It is generally low-lying and rises to about 600 m above sea level on the Kilungu hills. The

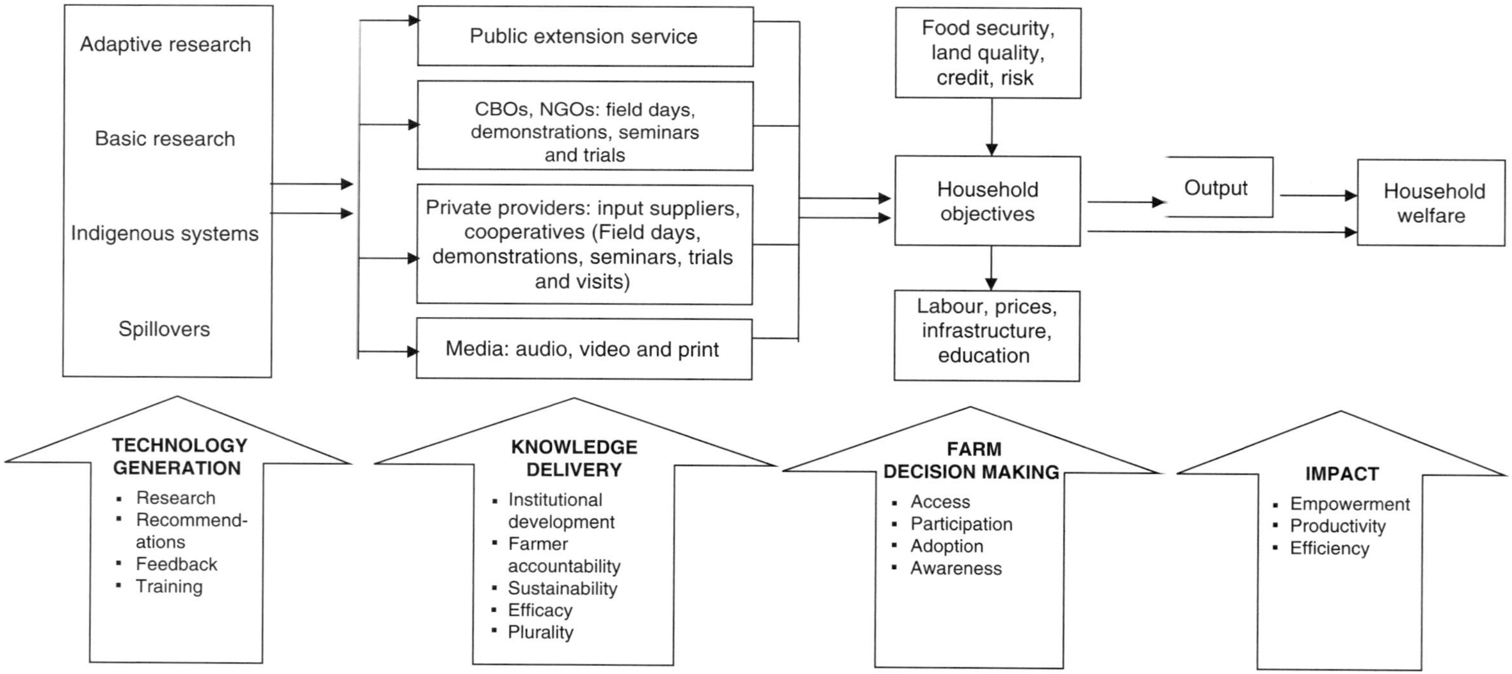


**Fig. 4.1.** Extension channels and pathways of impact on farmer household welfare. (Adapted from Gautam, 1999.)

major land features include the volcanic Chyulu hills and granite rocks forming Mbooni and Kilungu hills to the west. These hills receive generally good rainfall and thus support coffee, horticulture, livestock production and afforestation. The southern parts of the district comprise the low-lying grassland plains that have great potential for ranching. Rainfall in the south is scarcer and varies with altitude. The hilly masses to the north and central parts of the district are cool and wet, receiving 800–1000 mm rainfall per year, whereas the low-lying areas are hot and dry and receive 200–900 mm of rainfall per year.

According to the 1999 population census (CBS, 2001), the district had a population of 771,545. Livestock production is the major economic activity. Major livestock include cattle, sheep, goats and chickens. The main livestock products, in order of economic importance, are meat, hides and skins, honey, chickens and eggs. Crop farming is mainly for subsistence with maize, beans, pigeon peas and cowpeas as major food crops. Cash crops such as coffee and cotton are grown by a small proportion of farmers. Irrigated horticulture is also undertaken but largely by individual smallholder farmers.

This district was selected because information on the distribution of civil society institutions by divisions was available and showed considerable within-district variation (IPAR, 2000). Such information is crucial to a study of decentralization and CBOs. This information was overlaid with estimates of headcount poverty rates by divisions (Kenya, 1998), leading to selection of study divisions and locations that differed mainly by degree of distribution of civil society institutions.

Makueni district has 18 administrative divisions, 2 of which are wildlife conservation areas, while the rest are settled by farming communities. Based on available resources, two divisions, Kibwezi and Makindu, were purposively sampled based on a number of criteria. First, these neighbouring divisions are located in the southern part of the district and have similar communication networks, with their headquarters situated along the Nairobi–Mombasa road. So comparison between them minimizes differences due to agroecology, infrastructure or jurisdictional leadership or history. Second, they have similar agricultural activities, dominated by subsistence farming and livestock rearing (Kenya, 1997). Third, they have similar poverty situation, each with about 52% of households living below the poverty line (Kenya, 1998). However, Kibwezi has a much higher concentration of agriculture-based civil society institutions than Makindu does (Table 4.1).

The number of active agriculture-based civil society institutions was a stratification variable since they are vehicles for decentralization of extension service. NGOs and CBOs use dialogue, training and participatory development methodologies to bring community interests and problems to light, and can often better enable communities to address their problems (Owiti, 2002). This is critical as poor communities often lack the resources and leverage to influence decision making, especially at higher levels of development policy design. The intended impact of civil society organizations in aspects of resource use is to promote empowerment of individuals and communities (Uphoff and Wijayaratna, 2000; Pretty and Ward, 2001).

**Table 4.1.** Information on extent of civil society participation in study divisions. (From Kenya, 1997; IPAR, 2000.)

| Organization | Type | Target group | Present in Kibwezi | Present in Makindu |
|---|---|---|---|---|
| GAA | Donor organization | Farmers and groups | Yes | Yes |
| ARIDSAK | Donor organization | Farmers and groups | Yes | No |
| Care Kenya | NGO | Farmer production groups | Yes | No |
| Action Aid | NGO | Farmers and schools | Yes | No |
| AMREF | NGO | Farmers and school children | Yes | Yes |
| MACOSUD | CBO | Farmers and schools | Yes | No |
| KEC | CBO | Farmers and youths | Yes | No |
| KIP | Donor/GoK organization | Farmers and groups | Yes | No |

Information on the activities of CBOs and NGOs was obtained from the District Social Development Officer (DSDO) at the district headquarters in Wote. Kibwezi division, with an area of 945 km$^2$ and 16,282 households, had eight active NGOs and CBOs dealing with agricultural information dissemination. Makindu is 880 km$^2$ with 9907 households and two active NGOs and CBOs. Therefore, Kibwezi presented an area of higher extension decentralization than Makindu. As already indicated, there are relatively few and modest differences between the two divisions; thus, it is difficult to decipher why there are more agriculture-based civil society institutions in Kibwezi than in Makindu.

In addition to the NGOs and CBOs registered at the DSDO's office, there existed a number of other non-registered agriculture-based CBOs and self-help groups operating in the district which liaised with the registered (and funded) ones to have greater outreach coverage. The DSDO did not have a record of such organizations, as they did not liaise with them in implementation of their activities. Information on participation in such organizations was thus collected during the household survey.

Because of the vastness of the selected divisions, two administrative locations were randomly sampled from each division for data collection. Both divisions have four locations. The sampled locations were Kiboko and Twadu in Makindu, and Kikumbulyu and Utithi in Kibwezi.

Households were then randomly sampled from the selected location using a sampling frame obtained from the Central Bureau Statistics (CBS, 2001). There are 144,320 households in Makueni district, of which approximately 80% engages in agriculture. Using the statistically accepted procedure for selecting a sample from a given population ($\alpha = 0.05$), a minimum sample size of 83 would have been suitable to make statistical inference about the population. However, this study sampled about 125 households, which was dictated largely by resource availability, but taking into consideration likely instances of non-response so as to maintain a statistically valid sample size.

## 2.2 Data

Both primary and secondary data were collected from farmers and other stakeholders in Makueni district through literature review, informal surveys and formal surveys. Primary data for this study were collected using both qualitative and quantitative methods. The qualitative data collection involved focus group discussions and key-informant interviews aimed at obtaining first-hand perceptions about the effectiveness of decentralization in facilitating community participation, empowerment and local accountability, especially with regard to access, and utilization of agricultural extension information. The key informants interviewed were:

1. District officials, including District Veterinary Officer (DVO), District Agricultural Officer (DAO), District Agricultural and Livestock Extension Officer (DALEO) and the District Social Development Officer (DSDO);
2. Personnel from district-based NGOs dealing with agriculture;
3. Leaders from CBOs dealing with agriculture;
4. Personnel from regional research centres within the district;
5. Officials of other organizations deemed relevant, like regional offices of drug companies.

Following this qualitative research, a questionnaire was developed and pre-tested for data collection with a group of trained enumerators. Data were collected on:

1. Access to and utilization of extension services;
2. Sources of extension service – respondents were asked questions about advantages and disadvantages of the main sources of extension service, and also asked to rate each source of extension service based on costs, quality and accessibility;
3. Household characteristics (age, education level, gender of household head, etc.);
4. Household income and wealth;
5. Participation in farmer organizations or agriculture-related CBOs by household members.

Most of these variables are straightforward, although one merits more detailed explanation.

Farmers' wealth status was derived from a method suggested by Ahuja *et al.* (2004). In deriving this variable, the survey instrument required respondents to give the number of houses on the farm built with different construction materials (e.g. brick walls, tin roof, grass-thatched roof, etc.), the number of exotic grade and local cattle, grade and local chicken, as well as sheep and goats. An asset index was constructed from these variables using the following formula:

$$A_i = \sum_k f_k \frac{a_{ik} - a_k}{s_k} \qquad (4.1)$$

where $A_i$ is the value of the index for the $i$th household, $f_k$ is the factor score coefficient for the $k$th asset obtained from principal components analysis, $a_{ik}$ is the value of the $k$th asset for the $i$th household, and $a_k$ and $s_k$ are mean and standard deviation of the $k$th asset over all households.

The wealth status was constructed as follows: if the asset index was less than the mean, the wealth status was designated as 'poor'; if the asset index was between the sum of the mean plus one standard deviation, the wealth status was designated as 'middle class', and if the asset index was greater than one standard deviation above the mean, the household was designated as 'rich'. Data on derivation of the asset index are presented in the Appendix.

Summary statistics of farm and farmer characteristics appear in Table 4.2. The ratio of male- to female-headed households was about 3:1. There were no statistically significant differences with regard to age of household head, gender, education level of household head, household wealth status and household size between the two divisions. But farm and herd size were significantly larger in Makindu while household income was higher and access to telephones greater in Kibwezi.

# 3 Empirical Results and Discussion

## 3.1 Farmers' CBO membership and knowledge of extension delivery channels

Membership in local CBOs and knowledge of different extension service delivery channels were significantly higher in Kibwezi division. To obtain information about farmers' knowledge (awareness) of different channels, they were asked to list all the sources of extension services that were available in their locality within a specified period,[1] regardless of whether they utilized the source or not.

Knowledge and participation are both prospective indicators of empowerment within the community. In at least some places, farmer organizations have been shown to greatly improve the efficiency and provision of extension service in production activities (Uphoff and Wijayaratna, 2000). Social capital, as reflected in CBO membership, has also been shown to increase accountability of extension service to clients. For example, Vanessa (1997) noted that farmers, through their organizations, can be active partners in extension and can set the agenda and direct the process by which NGOs and government agencies participate to fulfil their needs and those of the farmers and the community. Uphoff (1996) further observed that introducing farmer organizations to irrigation systems in Gal Oya, Sri Lanka, improved resource utilization, enhanced access to extension and led to production of rice with minimal water quantities than would have been otherwise possible.

[1] The period between the last general election (parliamentary and presidential) held and the interview date was used as the reference period.

**Table 4.2.** Summary statistics for farmers' characteristics.

| | Decentralization level | | |
|---|---|---|---|
| | High (Kibwezi) (*n* = 60) | Low (Makindu) (*n* = 65) | Pooled (*n* =125) |
| **Frequencies**[a] | | | |
| *Sex* | | | |
| Males | 43 (71.7) | 49 (75.4) | 92 (73.6) |
| Females | 17 (28.3) | 16 (24.6) | 33 (26.4) |
| *Formal education* | | | |
| No formal education | 10 (16.7) | 18 (27.7) | 28 (22.4) |
| Primary level | 17 (28.3) | 25 (38.5) | 42 (33.6) |
| Secondary level | 14 (23.3) | 15 (15.1) | 29 (23.2) |
| Post secondary level | 19 (31.7) | 7 (13.5) | 26 (20.8) |
| *Wealth class* | | | |
| Poor | 36 (60.0) | 38 (58.5) | 74 (59.2) |
| Middle class | 14 (23.3) | 19 (29.2) | 33 (26.4) |
| Rich | 10 (16.7) | 8 (12.3) | 18 (14.4) |
| Access to telephone | 20 (33.3)[c] | 4 (6.2)[c] | 24 (19.2) |
| Access to radio | 57 (95.0) | 63 (96.9) | 120 (96.0) |
| Access to television | 7 (11.7) | 3 (4.6) | 10 (8.0) |
| **Means**[b] | | | |
| Age of household head (years) | 52.1 (1.73) | 49.2 (1.73) | 50.59 (1.08) |
| Household size | 6.9 (0.460) | 7.6 (0.60) | 7.3 (0.380) |
| Land size (acres) | 6.98[d] (0.964) | 14.81[d] (2.30) | 11.02 (1.32) |
| Cropped area (acres) | 4.08 (0.43) | 3.98 (0.48) | 4.03 (0.32) |
| TLU | 3.09[d] (0.484) | 4.70[d] (0.611) | 3.93 (0.398) |
| Income (KSh) | 11,782.5[d] (1,749.2) | 6,196.9[d] (851.0) | 8,899.6 (982.3) |
| Nearest Agrovet (km) | 5.61 (0.660) | 5.03 (0.510) | 5.31 (0.411) |
| Nearest AHA (km) | 7.93[d] (0.715) | 5.20[d] (0.512) | 6.51 (0.450) |
| Nearest market (km) | 2.53[d] (0.265) | 4.01[d] (0.389) | 3.30 (0.247) |
| Membership in local CBOs (number) | 1.45[d] (0.115) | 1.08[d] (0.119) | 1.26 (0.083) |
| Known extension delivery channels (number) | 5.26[d] (0.306) | 2.89[d] (0.208) | 4.03 (0.211) |

[a]Figures in parentheses in this section represent percentages.
[b]Figures in parentheses in this section represent standard errors.
[c,d]Figures with the same superscript are significantly different ($P < 0.05$).

Higher levels of social capital development and improved knowledge of extension sourcing are thus unconditionally associated with decentralization, although it is effectively impossible to establish the direction of causality, if any, in this relation, especially in simple cross-sectional data. However, evidence collected from the survey showed that most of the farmer associations were formed after NGOs and CBOs had initiated decentralized extension activities.

**Table 4.3.** Farmers' knowledge of different channels for extension service delivery.

| | Decentralization level | | |
|---|---|---|---|
| | High ($n$ = 60) Kibwezi | Low ($n$ = 65) Makindu | Pooled ($n$ = 125) |
| **Channel mentioned (frequencies)** | | | |
| Agrovets | 43.3 | 40.0 | 41.6 |
| CBOs | 60.0[a] | 13.8[a] | 36.0 |
| Chiefs' meetings (*Baraza*) | 65.0[a] | 24.6[a] | 44.0 |
| Farmer-to-farmer | 55.0 | 66.2 | 60.8 |
| Field days | 35.0[a] | 6.2[a] | 20.0 |
| Government extension | 78.3[a] | 40.0[a] | 58.4 |
| Media | 30.0[a] | 4.6[a] | 16.8 |
| NGOs | 58.3[a] | 18.5[a] | 37.6 |
| Research stations | 21.7[a] | 41.5[a] | 32.0 |
| Seminars | 45.0[a] | 16.9[a] | 30.4 |

[a]Figures with the same superscript are significantly different ($P < 0.05$).

Furthermore, other factors that might increase demand and/or raise awareness about need for extension sourcing (i.e. education level, wealth status) were not significantly different between the two divisions. So it seems that decentralization is contributing in some measure to the enrichment of social capital and farmer knowledge of extension options in this area of Makueni district.

Information on the level of awareness of specific channels of agricultural extension service delivery is given in Table 4.3. Apart from private-sector-oriented delivery channels related to veterinary and agrofarm input shops (agrovets), knowledge of other forms of delivery channels was significantly greater in the region of higher-level decentralization.

In the administrative region with less decentralized extension provision, the functioning of the extension system appears not to have changed significantly, still depending on the public delivery approach. With this approach, extension cannot be expected to reach every farmer; hence the need for selectivity and reliance on farmer-to-farmer dissemination. This could be the reason why farmers were most aware of this approach in Makindu, the low-decentralization region. However, with greater decentralization, the focus seems to have shifted more to the use of contact groups (farmer organizations) as points of regular and systematic interaction with the farming community, leading to a higher level of awareness of extension sourcing. Despite most (96%) farmers owning a radio, only a limited number demonstrated awareness of media as a form of agricultural information delivery.

## 3.2 Quality, cost and accessibility of agricultural extension delivery systems

Although agricultural information is commonly seen as a public good because of its non-rivalrous nature and limited excludability, like many other public goods, it can none the less be provided through private channels as well as public ones. Private extension can be categorized as the provision of a service or advice by a private-sector firm in exchange for a fee, the terms and conditions of which are negotiated in a competitive market (Chapman and Tripp, 2003). However, the degree to which this can be done in practice depends on the extent to which extension services can be converted into a private good. Specific examples of agricultural information that have private good characteristics could be tied to the use of a purchased input or service (like instruction on the use of a particular chemical), or client-specific advice catering to the specific needs of an individual farmer or group of farmers. Such information can be provided through agrovets, private veterinary practitioners or government extension officers acting in their private capacity. The main form of private extension service provision identified in the study region was through agrovets. Extension information that is difficult to privatize and commercialize includes time-sensitive production, marketing and management information of wide applicability. This type of information is typically best delivered through public systems, community-based systems or the mass media (e.g. radio programmes).

These various information delivery systems were all identified in the study region, operating through a variety of channels: (i) private systems, mainly agrovet shops; (ii) print, electronic and audio media; (iii) training, seminars and field tours organized by NGOs and CBOs, and to some extent farmer-to-farmer extension and home visits by CBO agents; (iv) public extension services via chiefs' barazas, Agricultural Society of Kenya (ASK) shows, research stations and visits to or from government extension agents. Farmers ranked the specific channels based on accessibility (i.e. whether the information is available when needed); quality (i.e. if the information is provided by well-trained/experienced personnel and is useful); and cost and affordability (i.e. the costs incurred in accessing the information). Each form was given a rank between 1 (best) and 4 (worst) for each descriptor. Farmers also provided an overall ranking of each delivery system. The results of the analysis are given in Table 4.4.

Public information delivery forms were ranked highest in terms of accessibility and quality, and in the overall preference ranking. Community-based systems ranked first in terms of cost, but second in accessibility, quality and overall preference rating. This should, however, be interpreted with care since

**Table 4.4.** Farmers' ranking of different extension delivery systems.

| | Ranking by the descriptor | | | |
|---|---|---|---|---|
| | Accessibility | Cost/Affordability | Quality | Overall rating |
| **Form of extension** | | | | |
| Public | 1 (47.2) | 2 (38.4) | 1 (51.2) | 1 (51.2) |
| Community-based | 2 (30.4) | 1 (44.0) | 2 (24.0) | 2 (27.2) |
| Private | 3 (7.2) | 3 (4.0) | 3 (11.2) | 3 (4.8) |
| Media | 4 (2.4) | 4 (0.8) | 4 (0.8) | 4 (3.2) |

Figures in parentheses represent percentages. A total of 16 farmers (12.8% of the respondents) were unable to satisfactorily respond to this question.

information from key-informant interviews revealed that, in some circumstances, NGOs did not employ their own staff, but facilitated public officers to offer services to farmers. The extent to which this facilitation contributed to public systems being ranked first could not be established in this study.

## 3.3 Factors influencing access to extension services

The question of greatest interest is who accesses extension services and whether this is influenced at all by decentralization of services provision and membership in CBOs. Towards that end, data were generated by the formal survey and those from the DSDO on the concentration of civil society institutions in each division. A multivariate analysis was undertaken to assess the relationship between decentralization, membership in agriculture-based CBOs and utilization of extension services. Differences across divisions in utilization of extension services cannot be attributed exclusively to differences in the extent of decentralization, since other unmeasured characteristics may be correlated with both divisions and extension services access. None the less, considerable attempt is made to control for as many such differences as possible in the study design and through household-level survey data collection. Thus, a dummy variable is used for Kibwezi division, the one with greater decentralization, as an admittedly imperfect proxy for the extent of local decentralization of services provision. The following logit model was estimated:

$$c_{ij} = W_{ij}\alpha + X_{ij}\beta + Y_{ij}\gamma + Z_j\delta + \varepsilon_{ij} \tag{4.2}$$

where $c_{ij}$ is a dichotomous variable that takes the value of 1 if the household $i$, in division $j$, accessed extension service during the 1 year preceding the survey; $W$ denotes a vector of household characteristics (age, education level and gender of household head, wealth status, and the number of community organizations to which household members resident on the farm belong); $X$ denotes a vector of farm characteristics (total livestock units and total cropped area); $Y$ denotes the approximate farm income; $Z_j$ denotes division $j$, i.e. level of decentralization of extension service; and $\varepsilon_{ij}$ is a standard regression error term.

The variable $c_i$ in Equation 4.2 was disaggregated further into informal and formal extension, described by Semana (1999). Informal extension has no syllabus or classroom. Rather it is focused on the farmer's problems and needs and is provided in the farmer's home or farm, or any convenient place. Formal extension, on the other hand, is planned, has written objectives and content that can be examined (but in most cases is not) and is typically carried out through short courses, field visits or tours for 1 or 2 days at community centres, research stations or county headquarters, or for a longer duration at designated farmer-training centres.

Informal extension was further divided into demand-driven vs supply-driven extension. This distinction is based on whether or not the farmer had requested extension assistance. If so, the access is considered demand-driven, while if the extension agent visited the farmer without the latter's request, it is deemed as supply-driven. Equation 4.2 was therefore estimated for three different types of extension service access: demand-driven informal extension, supply-driven informal extension and formal extension. Data did not support estimation of demand-driven formal extension.

In principle, estimation of the parameters in Equation 4.2 requires use of Heckman's two-step procedure to correct for selectivity problems associated with respondents actually demanding extension services during the reference period. However, since the whole sample fell in the 'expressed-demand' category, there is no variation in this dimension and thus no need to correct for potential selection bias. A similar approach has been used to evaluate the impact of availability and density of health infrastructure on access to health care in Peru (Valdivia, 2002).

Descriptive statistics for the variables used in the estimation of the three models are given in Table 4.5. Of the 125 farmers interviewed, 43 (34.4%) accessed demand-led informal extension services in the year preceding the survey; 22 (17.6%) received supply-led informal extension visits from agents; and 27.2% attended formal extension training. Formal extension training consisted of seminars (17.6%), tours to demonstration sites (5.6%) and field days (4.0%) of the sample. All formal extension activities were organized by NGOs and CBOs either individually or in collaboration with the public-sector extension service.

The logit model estimation results (Table 4.6) demonstrate the intuitive responsiveness of farmers' access to demand-led agricultural extension information to four factors, i.e. membership in CBOs, decentralization of extension services in the farmer's division of residence, income level and educational attainment.

Holding the other determinants constant, farmers in an area of higher decentralization were nearly three times more likely to access demand-led informal extension. This can be attributed to the fact that NGOs initiate collaboration with government agencies or with community organizations so as to scale up the success of their programmes and reach a larger number of persons within the community. Conversely, government agencies seek collaboration with NGOs so as to enable a wider reach of their activities. Decentralization had no statistically significant effect on access to supply-led or formal extension, reinforcing the interpretation that this effect works through empowerment of farmers to pressure extension services to respond to their needs when extension responsibility is decentralized.

This interpretation is further reinforced by the independent impact of CBO membership on access to demand-led informal extension only, with no statistically significant impact on access to supply-led informal or formal extension services. Farmer organizations and other CBOs provide vehicles through which farmers can become actively involved in the planning and management of extension and can demand to receive relevant services that meet their needs. The odds ratio indicates that a farmer with active membership in two agriculture-based CBOs is nearly five times more likely to obtain access to extension services than a farmer who belongs to none.

There is, however, differentiated access to extension services even within the more decentralized division and among those belonging to CBOs. While

**Table 4.5.** Description of variables.

| Variable | Description | Sample mean | Standard deviation |
|---|---|---|---|
| EXDEMAND | Received demand-led informal extension within last 1 year: 1 = farmer accessed the service; 0 = otherwise | 0.344 | 0.48 |
| EXSUPPLY | Received supply-led informal extension within last 1 year: 1 = farmer visited; 0 = otherwise | 0.176 | 0.40 |
| EXTRAIN | Received formal extension training during last 1 year: 1 = farmer attended training; 0 = otherwise | 0.272 | 0.45 |
| AGE | Age (years) of the household head | 50.59 | 12.07 |
| SEX | Sex of the household head: 0 = female; 1 = male | 0.74 | 0.44 |
| DIVISION | Division of study: 1 = Kibwezi, 0 = Makindu, coded as dummy variable | 0.48 | 0.50 |
| CBONO | Number of CBOs in which family members resident on the farm actively participate | 1.26 | 0.94 |
| EDUC | Highest level of formal education attained by the household head. Coded as a categorical variable: 0 = no formal education; 1 = primary level; 2 = secondary level; 3 = postsecondary education | 2.42 | 1.06 |
| TLU[a] | Number of livestock owned (tropical livestock units: 1TLU = 250 kg): bull = 0.7; cow = 0.6; steer = 0.5; heifer = 0.4; calf = 0.2; shoat = 0.1; lamb or kid = 0.03 | 3.93 | 4.45 |
| CA | Total area in acres under cropping | 8.06 | 7.10 |
| INCOME | Income per month (KSh)[b] | 8899.60 | 10938.80 |
| WTHSTAT | Wealth status.[c] Coded as a categorical variable: 1 = poor; 2 = middle class; 3 = rich | 1.55 | 0.73 |
| MARKET | Distance of the homestead to the nearest market (km) | 3.30 | 2.76 |
| HQUATERS | Distance of the homestead to the extension service source (divisional headquarters) (km) | 6.51 | 5.03 |
| PHONE | Farmer's access to a phone: 1 = access; 0 = no access | 0.91 | 0.40 |

[a]Based on figures in ILCA (1990).
[b]Generated as the aggregate of income from sales of farm produce and off-farm income of household members resident on the farm.
[c]The wealth status was constructed from the asset index (AI), such that if AI was less than the mean, the wealth status was designated as 'poor'; if the AI was between the sum of the mean plus one standard deviation, the wealth status was designated as 'middle class'; and if the AI was greater than this sum, the household was designated as 'rich'.

these improve access, they do not fully level the playing field for smallholder producers.

In particular, income had a positive and statistically significant effect on access to demand-led informal extension services (and a positive, but statistically insignificant, effect on access to other forms of extension). This may be because income enables farmers to meet the transaction costs involved in obtaining services, including travelling to the point of residence of the extension agent or making a telephone call to such an agent. Access to a telephone substantially increases the odds of extension access of all forms, although these estimates are statistically imprecise, most likely because telephone access is so strongly cor-

**Table 4.6.** Logit estimates of access to agricultural extension service.

| Independent variables | Demand-led informal | | Supply-led informal | | Formal training | |
|---|---|---|---|---|---|---|
| | β | Exp (β) | β | Exp (β) | β | Exp (β) |
| AGE | −0.004 (0.614) | 0.996 | 0.034 (0.031) | 1.035 | 0.038 (0.025) | 1.039 |
| SEX (1) | 0.520 (0.712) | 1.681 | −2.289 (1.126) | 0.775 | −0.029 (0.732) | 0.972 |
| DIVISION (1) | 1.596*** (0.614) | 4.931 | 0.979 (0.718) | 2.662 | 0.381 (0.546) | 1.463 |
| CBONO | 0.892*** (0.307) | 2.441 | 0.133 (0.326) | 1.142 | 0.242 (0.263) | 1.274 |
| EDUC | | | | | | |
| EDUC (1) | −0.936 (1.122) | 0.392 | −2.346* (1.408) | 0.096 | −0.473 (1.099) | 0.623 |
| EDUC (2) | −0.024 (0.752) | 0.977 | −0.435 (0.851) | 0.647 | −0.407 (0.793) | 0.666 |
| EDUC (3) | 0.288 (0.722) | 1.334 | −0.483 (0.905) | 0.617 | 0.476 (0.724) | 1.609 |
| TLU** | 0.043 (0.065) | 1.044 | −0.114 (0.105) | 0.892 | 0.039 (0.063) | 1.040 |
| CA | 0.047 (0.073) | 1.048 | 0.017 (0.103) | 1.017 | 0.046 (0.068) | 1.048 |
| INCOME | 7.09E−5** (2.7E−5) | 1.000 | 3.15E−05 (3.53E−05) | 1.000 | 4.54E−05 (2.73E−05) | 1.000 |
| WTHSTAT | | | | | | |
| WTHSTAT (1) | 1.288 (0.984) | 3.625 | −1.767* (1.051) | 0.171 | −0.492 (0.805) | 0.611 |
| WTHSTAT (2) | 1.342 (0.980) | 3.825 | 0.798 (0.992) | 2.222 | −0.709 (0.832) | 0.492 |
| HQUATERS | −0.073** (0.042) | 0.929 | −0.146** (0.073) | 0.864 | 0.024 (0.035) | 1.024 |
| PHONE (1) | 1.306* (0.796) | 3.692 | 2.289 (1.496) | 9.865 | 1.307* (0.810) | 3.694 |
| Constant | 5.247*** (1.825) | 0.005 | −3.876* (2.346) | 0.021 | −4.752*** (1.735) | 0.009 |
| Number of observations | 124 | | 124 | | 124 | |
| Model chi-square | 30.661*** | | 32.097 | | 18.665 | |

Figures in parentheses represent standard errors.
*Significant at $\alpha = 10\%$; **Significant at $\alpha = 5\%$; ***Significant at $\alpha = 1\%$.
Exp. value used in computation of the odds ratio values.

related with income, wealth status and educational attainment that it is difficult in such a small sample to isolate the effect of telephone access itself.

With regard to educational attainment, there was no statistically significant difference in extension access of any sort between farmers of different levels of education. However, farmers who have attained below secondary education appear to be disadvantaged in terms of accessing demand-led extension, although this does not reach significant levels.

Since most of the supply-led extension visits were made by agents from the public sector, it is not surprising that they were biased towards wealthier farmers and those who had an education level beyond primary schooling. Poor farmers were less likely to receive a visit by an extension worker than were rich and middle-income farmers. The result that extension visits are influenced by income and education is not new. This had been identified earlier as one of the problems with the training and visit (T&V) approach to extension. Better-off farmers, those who could read and write and those who could afford new technologies were favoured against the poorer farmers (Gautam, 1999).

Unlike the case for demand-driven and supply-driven informal extension services, the chances of being selected for training increased statistically significantly with distance from the divisional headquarters. It may be intuitively concluded that the reason behind selection for training is to choose those who are far away from the operating offices of the extension staff, perhaps so as to increase the reach of extension information.

## 4 Conclusions and Policy Recommendations

Lack of proper extension services is partially to blame for poverty, according to poverty assessments conducted in ten districts in Kenya in 2000 (Kenya, 2001a). With current government reductions in spending, it is unlikely that these public services will be increased in the near future. This makes it all the more important to organize extension provision effectively so as to maximize farmers' benefit, given scarce financial resources.

One approach through which this can be realized is by decentralizing extension services. This will involve reorganizing government services delivery by transferring responsibilities to other institutions, giving rise to a pluralistic service delivery system. The overall objective will be increased relevance and responsiveness of these services to users, and improved access and participation by the users.

This study has found that by controlling for lots of farmer-level covariates, access to demand-driven, informal extension services is substantially and statistically increased in the more decentralized localities. However, for a pluralistic extension system to work, more will need to be done in terms of coordination, probably at district level between the various provider groups. Although this study shows partnership and synergy between groups including the government and communities, there appears to be little effective coordination of all groups at the district level. For instance, although some NGOs employ the services of government extension personnel and deal directly with farmer organizations, this information is not readily available either at the District Social Development or District Agricultural and Livestock Extension Offices. The government departments involved in extension should therefore work towards developing and operationalizing a strong institutional framework that will guide partnerships with other development agencies so as to reach rural areas. Entrenching this as part of the extension policy guide might be a good step in this direction.

Despite the emerging pluralistic system, this does not mean that the government should diminish its role in extension. There remains great need for government extension, for a variety of reasons. First, the government still has a responsibility to help its departments attain its vision of increased production, incomes and improved standards of living by 2015 (Kenya, 2001b). Furthermore, the government has the personnel, technical backup and infrastructure to implement broad rural development programmes. The private sector, NGOs and CBOs each have a different philosophy, approach and funding.

It therefore remains the responsibility of the public sector to advocate and intervene in areas in which other players have little interest.

Because of the changing role of the public sector and the philosophical differences between the various players, it is important to empower farmers' groups to ensure that they have the capacity to press effectively for the services they need. Farmers groups are an important means of obtaining a voice for communities with which to demand services and link up to extension and research. Extension policies should therefore put more emphasis on farmers' groups as facilitating organizations and should perhaps even help farmers organize themselves into CBOs.

These results do support the claim that CBOs and decentralization enhance access to extension services. The findings of this study are not sufficient to advocate for community-based extension, as has been done by others (e.g. Raussen *et al.*, 2001; Wambugu *et al.*, 2001), because little is known about how this really works in practice. There needs to be closer examination of the mechanisms of group formation and farmer-to-farmer extension using common interest groups. For example, how do groups assist in disseminating information? What are the linkages between these groups? How can they better collaborate to bring about development? More research is required to better understand community-based extension as a tool for bringing about sustainable rural development.

## Appendix

**Table 4.A1.** Factor score coefficients used to construct the asset index.

| Variable | Factor score coefficient | Mean | Standard deviation |
|---|---|---|---|
| **Type of housing (number)** | | | |
| Mud grass | 0.57812 | 1.1520 | 1.3682 |
| Mud tin | 0.38568 | 0.1600 | 0.6648 |
| Mud iron | 0.63182 | 0.5280 | 0.9296 |
| Stone tin | 0.18939 | 0.2720 | 0.8169 |
| Stone iron | 0.30694 | 1.4240 | 1.5976 |
| Wood tin | −0.02648 | 0.0480 | 0.2146 |
| Wood iron | 0.08207 | 0.1360 | 0.7333 |
| **Livestock owned (number)** | | | |
| Grade cattle | 0.56352 | 0.0800 | 0.4135 |
| Local cattle | 0.31087 | 2.3120 | 3.4017 |
| Goats | 0.28170 | 11.744 | 14.499 |
| Sheep | 0.34172 | 1.9040 | 4.1102 |
| Grade chicken | 0.54093 | 2.4720 | 19.935 |
| Local chicken | 0.48307 | 15.200 | 13.123 |
| Asset index | | −0.02525 | 1.72311 |

## References


Ahuja, V. and Redmond, E. (2004) Livestock services and the poor. *Tropical Animal Health and Production* 36, 247–268.

Amudavi, D.M. (2003) Advancing a partnership model of extension to support the Kenya National Agriculture and Livestock Extension program (NALEP) in rural livelihood improvement. AIAEE 2003, Proceedings of the 19th Annual Conference, Raleigh, North Carolina, pp. 43–57.

Anderson, J. and Van Crowder, L. (2000) The present and future of public sector extension in Africa: contracting out or contracting in? *The Public Administration and Development* 20, 373–384.

Antholt, C.H. (1994) Getting ready for the twenty-first century: technical change and institutional modernization in agriculture. World Bank Technical Paper No. 217, Washington, DC.

CBS (Central Bureau Statistics) (2001) *Population Census*, Vol. 1. Government Printer, Nairobi, Kenya.

Chambers, R. (1993) *Challenging the Professions: Frontiers for Rural Development*. Intermediate Technology, London.

Chapman, R. and Tripp, R. (2003) Changing incentives for agricultural extension: a review of privatized extension in practice. Overseas Development Institute, Agricultural Research and Extension Network. Network Paper No. 132, July 2003. Westminster, UK.

Gautam, M. (1999) Agricultural extension: the Kenya experience. Operations Evaluations Department. Report No. 19523, 30 June 1999. The World Bank, Washington, DC.

Hangrave, B. (1999) The role of religious-based NGOs in the research and extension process in the arid and semi-arid lands of Kenya. Case study prepared as part of the IGAD/INTSORMIL/USAID-REDSO study. Agricultural Technology for the Semi-arid Africa Horn. Lincoln, Nebraska.

ILCA (International Livestock Centre for Africa) (1990) Working Paper 1. Livestock systems research manual, Vol. 1. International Livestock Centre for Africa, Addis Ababa, Ethiopia.

IPAR (2000) *Social and Institutional Mapping: Isiolo, Makueni, Muranga, Kilifi and Selected Urban Poverty Settlements in Mombasa*. IPAR, Nairobi, Kenya.

Kenya (1997) *Makueni District Development Plan*, 1997–2001. Government Printers, Nairobi, Kenya.

Kenya (Ministry of Planning) (1998) *Poverty in Kenya*, Vol. 1: *Incidence and Depth of Poverty*. Ministry of Planning, Government Printers, Nairobi, Kenya.

Kenya (2001a) *National Agricultural Extension Policy*. Ministry of Agriculture and Rural Development, Government Printers, Nairobi, Kenya.

Kenya (2001b) *Poverty Reduction Strategy Paper, 2001–2004*, Vol. 1. Ministry of Planning, Government Printers, Nairobi, Kenya.

Kenya (2001c) *National Agriculture and Livestock Extension Programme (NALEP), Implementation Framework*. Ministry of Agriculture and Rural Development, Nairobi, Kenya.

Owiti, J.M. (2002) National Agriculture and Livestock Extension Programme (NALEP), Final Report of Advocacy component Review. IPAR, Nairobi, Kenya.

Pretty, J. and Ward, H. (2001) Social capital and the environment. *World Development* 29(2), 209–227.

Raussen, T., Ebong, G. and Musiime, J. (2001) More effective natural resource management through democratically elected, decentralized government structures in Uganda. *Development in Practice* 11(4), 460–470.

Semana, A.R. (1999) Agricultural extension services at crossroads: present dilemma and possible solutions for future in Uganda. Available at: www.codesria.org/Links/Conferences/ifs/ Semana.pdf

Smith, L.D. (1997) Decentralization and rural development: the role of the public and private sector in the provision of agricultural services. Technical Consultation on Decentralization, FAO, Rome, 16–18 December 1997.

Uphoff, N. (1996) *Learning from Gal Oya: Possibilities for Participatory Development and Post Newtonian Social Science*. Intermediate Technology Publications, London.

Uphoff, N. and Wijayaratna, C.M. (2000) Demonstrated benefits from social capital: the productivity of farmer organisations in Gal Oya, Sri Lanka. *World Development* 28 (11), 1875–1890.

Valdivia, M. (2002) Public health infrastructure and equity in the utilization of outpatient health care services in Peru. *Health Policy and Planning* 17, 12–19.

Vanessa, S. (1997) *Farmer-led Extension: Concepts and Practice*. Intermediate Technology Publications, London.

Wambugu, C., Franzel, S., Tuwei, P. and Karanja, G. (2001) Scaling up the use of fodder shrubs in central Kenya. *Development in Practice* 11(4), 487–494.

Wanga, E. (1999) Key note address on new perspectives in rural extension. Regional Refresher International Course in Rural Extension (ICRE) on Challenges and Prospects, Egerton University, 21 November–3 December.

# 5 Access to, and Willingness to Pay for, Agricultural Extension in Western Kenya

E. NAMBIRO AND J.M. OMITI

## 1 Introduction

Agricultural extension aims to increase farm productivity and improve the welfare of rural people by educating farmers on advanced farming techniques and promoting an innovative environment (Butcher, 1984; Arnon, 1989; Evenson, 1998). Extension achieves this by linking researchers, government planners, non-governmental organizations (NGOs), community-based organizations (CBOs) and the private sector with farmers, and offering an open platform for the exchange of ideas and services (Farrington, 1995). As such, the role of agricultural extension workers is evolving to integrate farmers' knowledge with formal science and to build a culture of dialogue among various actors and partners (Millar, 1994). The functional shift is necessitated by the realization that sustainability is best achieved if farmers take on more active and participatory roles in extension provision than they previously have.

This increasingly dominant view of extension is consistent with the rising trend of decentralizing to more local levels the resource allocation and mobilization functions related to agricultural extension. In Kenya, extension services have been decentralized to the division level. Decentralization of agricultural extension services is aimed at improving farmers' access to these services. There are three main reasons why governments decentralize: (i) local governments are better placed to effectively handle complex, location-specific issues; (ii) central governments are unable to adequately finance and deliver a sufficient range of services; and (iii) the view that democracy can be best served through devolved functions (Smith, 1997). However, the more nebulous notions of popular participation and accountability have come to be regarded as the raison d'être of decentralization. Hence the growing tendency for the design of development programmes with in-built local participation and accountability in Kenya.

In Kenya, the current National Agriculture Extension Policy (NAEP) was drafted in 1994 and became operational in 2001. It aims to make the national extension system more responsive to the needs and realities of its farm-level clients and to promote linkages between the ministry and other stakeholders in agricultural extension. NAEP is implemented through the National Agriculture and Livestock Extension Programme (NALEP) (Kenya, 2001a). Under the NAEP, Kenya has decentralized agricultural extension. However, the success of agricultural extension decentralization hinges on the answers to two key questions. First, are decentralized extension services accessible to all farmers and are there observable differences in quality, cost and client confidence based on the method of extension delivery? Second, especially since government policy is moving towards increased pay-as-you-go (i.e. user fee–based) service provision, are farmers willing to pay for extension services? This study investigates these two questions in Kakamega district in western Kenya.

## 2 A Brief History of Agricultural Extension in Kenya

Strategies such as the Special Rural Development Programme (SRDP) and the District Focus for Rural Development (DFRD) of 1980 were designed with the intent to create a development policy that would improve the efficiency of extension services delivery in Kenya. The provision of agricultural extension is a matter of public interest given agriculture's significant contribution (26%) to gross domestic product (Kenya, 2001b) and the fact that it provides a livelihood for 70% of Kenya's rural population (Kenya, 2001a). The mode of extension provision has evolved alongside changes in development practice and has included individual farm visits, group approach, farmer field schools, unified extension, farm management, integrated development and specialized commodity extension over time (Evenson, 1998).

The Farming Systems Research and Extension (FSR/E) model, which was operationalized between 1965 and 1980 in response to concern for small-scale farmers, was characterized by farm-level participation through on-farm trials and three-way linkages between farmers, researchers and extensionists (Collinson, 2000). The most notable success of the FSR/E model was the dissemination of hybrid maize. However, the FSR/E approach often suffered from a mix of ad hoc project components, lacked a consistent national strategy, was high cost and exhibited a disproportionate focus on male farmers and large landowners. Due to these and other shortcomings, FSR/E was judged an inefficient and cost-ineffective model (Birkhaeuser *et al.*, 1991; Gautam, 1999).

In response, the training and visit (T&V) extension approach, which had earlier been used in Turkey and India, was introduced in Kenya in 1982 as part of the agricultural growth strategy funded by the World Bank and implemented under the National Extension Project (NEP). The major objective of NEP was to develop institutions that would facilitate delivery of agricultural extension services to smallholder farmers efficiently and effectively. NEP aimed to train competent, well-informed village-level extension workers who would visit farmers frequently and regularly with relevant technical messages and take farmers' problems to researchers. NEP was implemented in two phases: NEP I (1983–1991) focused on extension services for crop production, while NEP II (1991–1998) changed strategy to integrate crop and livestock production (Benor and Daxter, 1984; Kenya, 2001a).

The T&V model under NEP I and II succeeded in putting an integrated national agricultural extension system in place, improved the quality of staff through training, established better researcher–farmer linkages, raised public extension staff morale and reduced some of the previous biases against women, small-scale farmers and remote areas. However, with time, new biases cropped up in favour of more educated and more productive farmers. Other drawbacks were the top-down, supply-driven approach, the lack of focus on farmer empowerment, poor project implementation, weak financial management, poor development of the relevant key institutional features, and a lack of accountability and responsiveness from management that often led to a disparity between the advisory messages and farmers' needs (Evenson, 1998; Gautam, 1999).

All the above limitations arose because T&V's organized and strictly structured method of agricultural extension assumed the availability of a sustained flow of research innovations, the ability of implementing agencies to secure, retain and motivate good technical staff, and sufficient funds to continuously sustain the effort. However, these assumptions did not apply to many parts of Kenya, including Kakamega. Highly differentiated farming conditions rendered the research pipeline insufficient for many crops and agroecological conditions, and implementing organizations were often poorly managed, underskilled and underfunded.

Because of the many weakness of NEP, a new, farmer demand–driven extension approach, NALEP, was designed by the Ministry of Agriculture and Rural Development (MoARD) in collaboration with the Swedish Government in 1998 (Kenya, 2001a). In designing NALEP, the positive and negative aspects of the earlier approaches were considered. The change in extension models was partly precipitated by increasing recognition of the complex socio-economic and agroecological conditions faced by resource-poor farmers (Farrington, 1998). Research and extension agencies did not have the capacity to generate the appropriate mix of technologies required by farmers (Thrupp and Altieri, 2001). NALEP was designed to respond to such challenges based on a broader and more inclusive 'farmer-oriented' extension service that was better equipped to meet the needs and demands of the small-scale farming population, operationalized through the formation of focal area development committees and common interest groups (Kenya Rural Development Strategy and the Poverty Reduction Strategy). NALEP's focal area extension approach is intended to help government mobilize increased private-sector and other stakeholder participation in decision making and extension service provision (District Agriculture Officer, Kakamega, 2004, personal communication). Extension agents and other stakeholders spend 12 months in a focal area of about 400 households, intervening according to a community action plan derived in a participatory manner. The NALEP team also organizes farmers in the focal area into common interest groups (CIGs) for ease of training. Appropriate technical packages are then promoted based on CIG action plans and farm-specific action plans, with deliberate targeting of resource-poor farmers (J. Wangila and B. Swallow, 2001). NALEP thereby aims to support a pluralistic approach involving all key stakeholders and to facilitate a gradual transition from predominantly publicly supported extension to private provision of agricultural extension services.

However, the operational weaknesses in NEP that resulted in limited access and lack of new technical messages to farmers have not yet been fully addressed. This has led to some extension staff reverting to the old T&V style of disseminating messages through public administration meetings (barazas). Nevertheless, NALEP has sparked some innovative partnerships wherein alternative extension service providers, such as the private sector, NGOs, CBOs, faith-based organizations (FBOs) and others, work together to meet extension objectives, and reinforce the decentralization and privatization of extension services. The question is whether it is working and whether it is building farmer-level demand for extension services, manifest in client assessment of quality, cost and accessibility, as well as willingness to pay (WTP) for extension services. Especially because sluggish economic growth has forced cutbacks in public agricultural extension budgets (Gautam, 1999), fee-based extension services from either public agencies or private providers have attracted considerable interest of late.

# 3 Study Area, Data and Descriptive Statistics

## 3.1 Study area

Kakamega district is located in Western Province (Fig. 5.1) and comprises six divisions: Municipality, Shinyalu, Ikolomani, Lurambi, Kabras and Ileho. The district covers an area of 1498.5 km$^2$ with a population of 603,000 in 100,500 households, 70% of whom depend on agriculture for their livelihood, as per the 1999 census (Kenya, 2001c). The altitude ranges from 1400 to 1600 m above sea level. Kakamega was chosen as the study site because the implementing institution, the Institute of Policy Analysis and Research (IPAR), had previously undertaken several studies in the area and had developed good contacts with local organizations and persons critically involved in decisions and consultations about agricultural extension services.

## 3.2 Data collection methods

Structured questionnaires were used to collect farm-level data. These were supplemented by interviews with key informants drawn from government and NGOs. The aim of the key informant interviews was to obtain first-hand information on the role of farmer participation in extension provision in facilitating local accountability and empowerment, especially with regard to access to, and use of, agricultural extension information. Due to limited resources, purposive sampling was used to select Ikolomani (a rural division) and Municipality (an urban division) among the six divisions in Kakamega. Multistage random sampling was then used to select two locations from each division and two

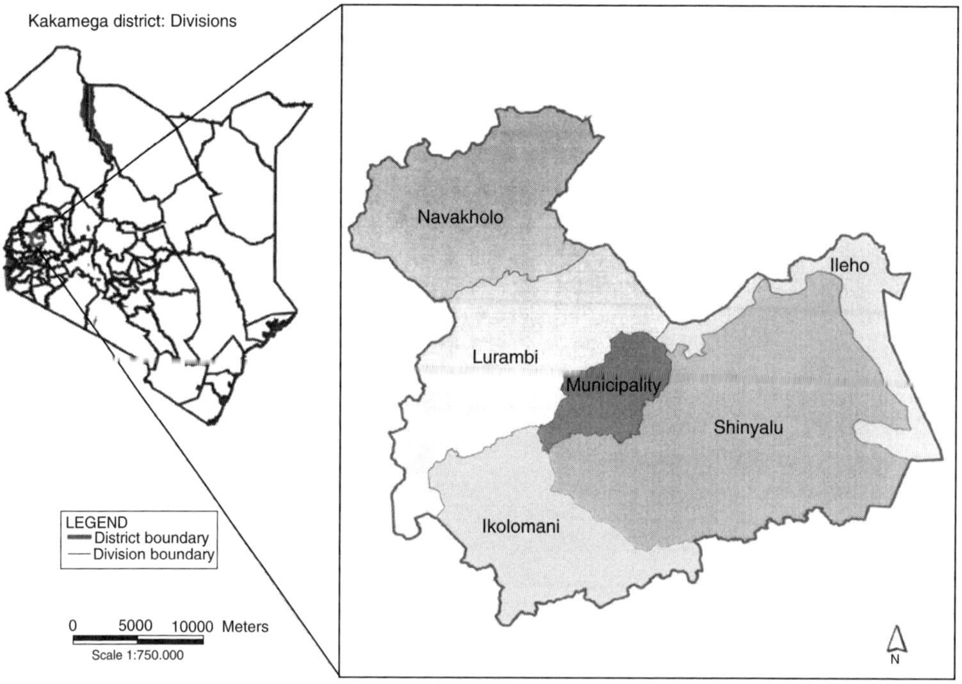


**Fig. 5.1.** Kakamega district.

sublocations from each location. A total of 112 farmers were randomly selected from the eight sublocations of the district (i.e. 14 farmers from each sublocation drawn from lists provided by the agricultural extension personnel) and interviewed individually in April 2004 about the preceding crop year. However, due to missing data only 110 questionnaires were used in the analysis. Data collected covered household characteristics, farm enterprises, use of inputs, postharvest handling activities and respondents' views on extension services. Additionally, data on extension services availability, distance from the nearest extension office, extension services supply and demand, and accountability of service providers to the local communities were also collected.

Table 5.1 presents basic sample descriptive statistics. The average household size was 6.4 persons. Of the household heads 63% were male and 50% had completed primary education. Within households, male members accounted for 49%. Over 90% of the households possessed a radio and could access information via radio and 40% had access to either telephone or television. About 61% relied solely on agriculture activity as their source of livelihood and the average land size was 0.91 ha per household.

### 3.3 Receipt of extension services by source

In the previous year, 41% (45 households) of farmers received agriculture extension information through various sources (Table 5.2). The government

**Table 5.1.** Household characteristics (*n* = 110).

| | | Percentage |
|---|---|---|
| Sex of the household head | | |
| Males | | 63 |
| Females | | 37 |
| Formal education of the household head | | |
| No formal education | | 9 |
| Primary education | | 34 |
| Secondary education | | 35 |
| Technical education | | 21 |
| Other | | 1 |
| Variable | Sample mean | Standard deviation |
| Land size in acres (ha) | 0.91 | 1.4 |
| Ownership of radio (0 = no, 1 = yes) | 0.94 | – |
| Ownership of television (0 = no, 1 = yes) | 0.41 | – |
| Ownership of telephone (0 = no, 1 = yes) | 0.30 | – |
| Distance from house to market (km) | 2.3 | 1.2 |
| Distance from house to government extension office (km) | 4.9 | 5.2 |
| Sex of household head (0 = female, 1 = male) | 0.63 | – |
| Age of household head (years) | 48.42 | 13.75 |
| Education level of household head (0 = no formal education, 1 = primary, 2 = secondary, 3 = tertiary/college, 4 = other) | 2.71 | 0.935 |
| Employment of household head (farmer is fully employed; 0 = no, 1 = yes) | 0.89 | – |
| Average off-farm monthly income of household (KSh) | 7348 | 5454.9 |
| Maize yield (90 kg bags) per hectare | 6.49 | 12.58 |
| Number of dairy cattle owned | 0.760 | 1.349 |

**Table 5.2.** Main sources of agricultural extension services (*n* = 45).

| Sources of agriculture extension services | Households citing source as most important (%) |
|---|---|
| Government extension worker | 22.0 |
| Media (radio, newspapers and television) | 16.9 |
| Neighbours/friends/relatives | 13.4 |
| Agrovets | 11.3 |
| Field days | 9.9 |
| Chiefs' barazas | 8.6 |
| Privately employed AHAs | 5.6 |
| Research stations | 6.2 |
| Seminar/training/tours | 3.5 |
| NGOs | 1.1 |
| CBOs | 0.9 |
| Hawkers of drugs, pesticides | 0.6 |

extension office was the most common source of extension advice (22.0%) followed by the media (16.9%).[1]

If we group delivery systems as private (agrovets and privately employed animal health assistants, AHAs), public (research institutions, government departments, agricultural shows of Kenya chief's barazas), mass media (print, electronic and audio), and CBOs or NGOs, 64% of those receiving extension services access them through public or mass media channels. Extension advice through or from government staff is the most common means of accessing extension services. Public animal health and husbandry extension services commonly require some payment for particular interventions, as determined by government regulation. Private, for-profit extension services are relatively expensive – for example, most veterinary service providers in the district charge a KSh 200 consultation fee per visit – while CBO or NGO systems have had little impact in the survey areas.

Only 29% of the respondents had sought and received extension advice on agricultural production or marketing since April 2003, indicating access to demand-driven extension services, while 12% had received extension services without asking for the service. We term this latter sort supply-driven extension service. Some farmers received multiple extension services. As much as 59% of the demanded services dealt with animal health, 18% with animal production and 13.3% with crop production. Costs incurred in order to get the extension services ranged from nothing for crop production services to KSh 6900 for veterinary services. The costs were zero especially when the extension was supply-driven from the extension agents and provided by government, NGOs, CBOs and occasionally from the agrovets when marketing their products. Supply-driven extension focused mainly on methods of good crop husbandry, animal health and farm record keeping.

## 3.4 Quality and affordability of extension services by source

Most extension information delivered to farmers concerns crop production and livestock husbandry practices. As a genuine public good (i.e. use of the information by one person in no way precludes its use by others), such information can be delivered most easily through mass means such as the radio, pamphlets and group meetings, rather than more expensive individual home visits. Unit costs of delivery, however, do not seem to match well with farmers' assessments of relative quality and affordability.

The four major extension and information delivery systems – private, public, media, and NGOs and CBOs – were compared and ranked by farmers in terms

[1] Media-delivered extension occurs through print, television or radio mostly prepared by the Ministry of Agriculture personnel or other recognized stakeholders. In the print media, different topics of agriculture technology development are tackled on particular days of the week in the daily newspapers. On the radio, programmes such as *Tembea na majira* are aired on a weekly basis by the Kenya Broadcasting Cooperation (KBC) that educate farmers on different crops and animal husbandry and recently introduced agriculture technologies.

**Table 5.3.** Extension services delivery channels rankings, by quality and affordability.

| Extension service delivery channel | Ranking by descriptors | |
|---|---|---|
| | Quality | Affordability |
| Public | 1 (49.5) | 1 (49.0) |
| Private | 2 (20.0) | 2 (19.8) |
| Media | 3 (16.3) | 3 (16.8) |
| Neighbours/friends/relatives | 4 (12.2) | 4 (12.0) |
| CBOs/NGOs | 5 (2.0) | 5 (2.4) |

Figures in parentheses are percent of respondents who ranked the delivery system as best.

of quality and affordability on a scale of 1–5, with 1 being 'best' and 5 'worst' (Table 5.3). In terms of both quality[2] and affordability, public extension delivery systems were ranked first by quite a large margin. By both standards, private providers came second, a bit ahead of mass media. NGOs and CBOs came last in rankings of quality and affordability of extension services provided.

# 4 Econometric Evidence

## 4.1 Factors that influence access to agricultural extension

As already shown, more than half the sample did not receive information from agricultural extension services in 2003. So what factors influence access to extension information? We estimate this using a probit regression model. The dependent variable was a dichotomous variable equal to 1 if extension service was accessed in 2003 or 0 if it was not.

Table 5.4 reports the probit estimation results. Households owning telephones and headed by older males had statistically significantly greater access to extension information. The introduction of mobile telephones in Kenya en masse in 1999 has revolutionized extension provision. Farmers who own mobile telephones are more likely to receive extension services promptly, presumably because they can more easily communicate with extension agents by phone than can households lacking a phone.

Male-headed households were more likely to access extension services, suggesting gender bias both in the choice of client that extension agents choose to visit and in the demand for extension services. The different traditional roles males and females play in the household may create cultural barriers reflected in the lower likelihood that a female head of household would demand extension advice. The male bias is further underscored as 30% of the male-headed households had attended some form of agricultural training or agricultural field day, compared with only 3% for female-headed households. Female farmers

[2] The criterion for quality was that the advice provided was relevant and timely, and added value to the client's operations.

**Table 5.4.** Probit results household receipt of extension services, 2003.

| Independent variables | |
|---|---|
| Household owns radio | 0.222 |
| Household owns television | 0.582 |
| Household owns telephone | 0.218* |
| Distance to government extension office | –0.01* |
| Male household head | 0.952** |
| Land size in hectares | 0.034 |
| Age of household head | 0.012* |
| Education level of household head | 0.260* |
| Average off-farm monthly household income | –0.004** |
| Constant | 1.5** |
| Number of observations | 110 |
| Pseudo $R^2$ | 0.46 |

*$p \leq 0.05$; **$p \leq 0.01$.

still appear to get ignored by agricultural extension despite decentralization of extension services in recent years.

Several other covariates help explain extension access. The probability that a farmer had received extension advice reduces as distance to the extension office increases, probably because of travel costs, including the opportunity cost of time. The educational level of the household head favourably influences farmers' access to extension information. The implied elite bias was a central criticism of previous (supply-driven) extension programmes based on T&V. These biases imply a poorly targeted system whereby less educated farmers who are likely to benefit the most from extension advice are largely ignored. Average off-farm income negatively and significantly affected probability that the farmer accessed extension services. This may be because people working off their farm are less often available on their farms during the extension officers' working hours. Overall, the low rate of access to extension services and the apparent continued gender and elite bias in extension delivery, especially traditional supply-driven extension, raises questions about the extent to which decentralization following the NALEP approach has succeeded in generating equitable and broad-based access to agricultural extension services.

## 4.2 Farmers' willingness to pay for extension services

One response to reductions in public service provision in some countries has been to commercialize extension services. However, commercialization of extension services is only possible to the extent that farmers are willing to pay for the services. WTP is defined as the maximum amount of money an individual would be willing to pay for a good or service rather than do without it. WTP for extension service is thus the amount of money that would make a farmer indifferent between paying for extension information services and foregoing extension support but retaining the money. Farmers' WTP for extension services can be evaluated through contingent valuation (CV) methods.[3] We elicit farmers' WTP by CV methods and investigate those data econometrically in this section.

Following Gorham (1998), we proposed various random levels of payment so as to elicit from respondents the maximum amount they were willing to pay for extension services per visit. Beyond describing the distribution of WTP for extension services, we can also relate the elicited household-specific WTP estimates for extension services to the covariates employed in the earlier probit regressions, but now using a Tobit model to account for the censored, non-negative nature of the WTP-dependent variable. Furthermore, in order to test whether past receipt of extension services is related to farmer WTP for extension access, we also test whether WTP in 2004 was higher among the farmers who received extension information services in 2003.

The WTP results suggest that there is a market for extension services in the district. Nearly half (49%) of the farmers were willing to pay for extension visits. One can conjecture that farmer payments for services will create better incentives for extension agents to deliver information that farmers want – i.e. to provide more demand-driven extension services – and that it might encourage entry of private extension service providers, thereby stimulating competition and improving coverage and quality. Figure 5.2 plots the frequency and cumulative frequency distributions of the subsample with positive WTP. The distribution is highly skewed, with a few people willing to pay relatively substantial sums (up to KSh 1000) for an extension visit, but a mode value of KSh 152.

These results imply that there is demand by farmers for decentralized extension information services, even if they must pay for the service. This corroborates the World Bank's findings (Gautam, 1999), which indicate that extension information delivery systems were among the services that farmers would pay for if improved. We do not study the costs of service provision, so we cannot establish whether these WTP levels would make extension provision commercially attractive. But it does suggest the very real possibility of incorporating some (at least partial) cost recovery mechanism (e.g. user fees) that could both generate resources locally to help expand the coverage and improve the quality of decentralized agricultural extension services, and that could help induce extension providers to emphasize more demand-driven information delivery.

So what factors determine WTP for extension services? Table 5.5 presents the results of a Tobit regression exploring the determinants of WTP for extension services. Again, owning a telephone is positively and significantly associated with a household's WTP. This may be because of improved communication between the extension agent and the farmer. Ownership of telephone also clearly signals relative wealth, so perhaps this merely captures the fact that willingness to pay is strongly associated with ability to pay. Unlike current patterns of extension service provision, however, there is no independent wealth bias (as manifest through the income or education variables) or gender bias in

[3] See Hanemann (1984), Cameron and James (1987) McFadden (1994) and Messonnier *et al.* (2000) for technical details on CV elicitation and estimation methods.

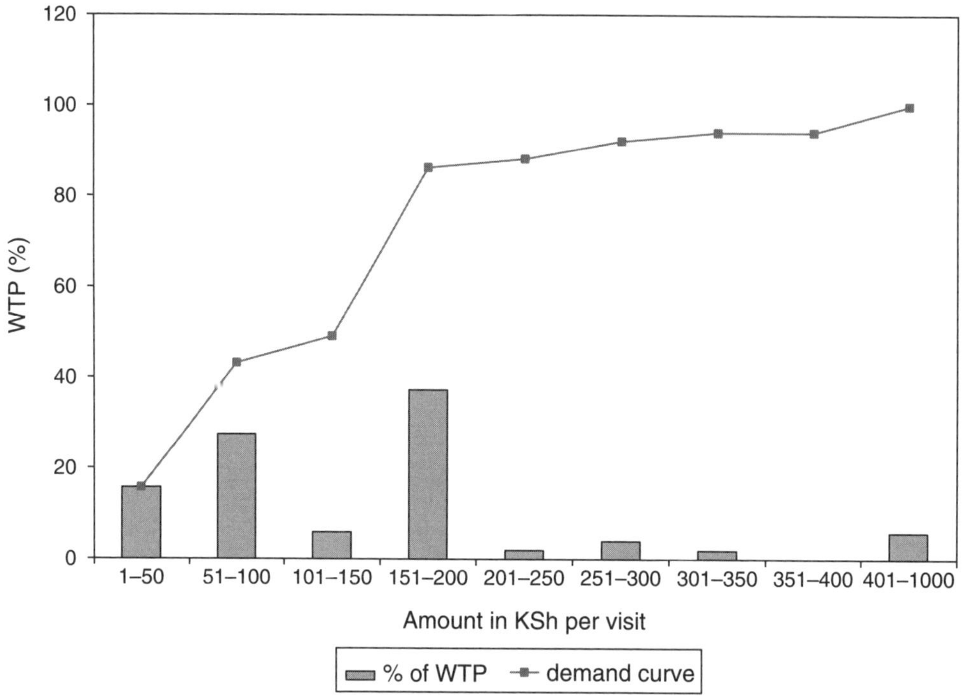


**Fig. 5.2.** Household willingness to pay for an extension visit.

**Table 5.5.** Tobit estimates of willingness to pay for extension services per visit.

| Independent variables | WTP (KSh) |
|---|---|
| Access to telephone | 77.25* |
| Distance to market | −4.11* |
| Male household head | −55.39 |
| Age of household head | −1.45 |
| Education level of household head | 0.91 |
| Average off-farm monthly income | 0.88 |
| Visited by extension in 2003 | 167.11** |
| Number of observations | 110 |
| $R^2$ | 0.489 |

*$p \leq 0.05$; **$p \leq 0.01$.

WTP for agricultural extension. However, those who had received extension services in the previous year were positively and significantly more willing to pay for the services. This may be attributable to satisfactory services received, to ongoing need for assistance or both. But it is clearly a positive sign that those with prior exposure to the services are willing to pay more for it.

## 5 Conclusions

This study of agricultural extension provision in Kakamega district shows that government extension service was the most accessible and the most highly regarded by farmers, in terms of both quality and affordability. Private-sector as well as CBO and NGO participation in extension service provision is still low and of uneven or low quality. Moreover, in spite of dramatic changes in government extension policy, much extension service delivery in the district continues to follow the long-standing, supply-driven model.

Regression results show that well-to-do and male-headed households were more likely to access extension information. No gender or wealth bias appears in analyses of farmers' willingness to pay for demand-driven extension services. This strongly suggests that further progress in the intended direction of demand-driven, decentralized extension service provision can reduce the apparent elite bias that has long plagued extension provision in Kenya, as in many other countries. Willingness to pay for extension services also showed that roughly half of all farmers were willing to pay an average of KSh 152 per extension visit. This raises the clear possibility of at least partial cost recovery by local government extension service providers, and possibly even commercially viable extension information delivery via local input suppliers (e.g. agrovets, veterinarians) and seed wholesalers.

## References

Arnon, I. (1989) *Agricultural Research and Technology Transfer*. Elsevier Science Publishers, New York.

Benor, D. and Daxter, M. (1984) *Training and Visit Extension*. World Bank, Washington, DC.

Birkhaeuser, D., Evenson, R.E. and Feder, G. (1991) The economic impact of agricultural extension: a review. *Economic Development and Cultural Change* 39(3), 507–521.

Butcher, C. (1984) Nomadic pastoralism and extension: a review of the literature. Pastoral Development Network, Paper 37d. Overseas Development Institute, London.

Cameron, T. and James, M. (1987) Efficient estimation methods for use with closed ended contingent valuation survey data: a review. *Economics and Statistics* 69(3), 269–276.

Collinson, M. (ed.) (2000) *A History of Farming Systems Research*. CABI Publishing and Food and Agriculture Organization of the United Nations, Wallingford, UK.

Evenson, R. (1998) Economic impact studies of agricultural research and extension. Yale University Economic Growth Center, New Haven, Connecticut.

Farrington, J. (1995) The changing public role in agricultural extension. *Food Policy* 20, 537–544.

Farrington, J. (1998) *Organization Roles in Farmer Participatory Research and Extension: Lessons from Last Decade*. ODI Natural Resource Perspectives 27. Overseas Development Institute, London.

Gautam, M. (1999) Agriculture extension: the Kenya experience. The World Bank Operations Evaluation Department, Report No. 19523, Washington, DC.

Gorham, E., DeVaney, S.A. and Bechman, J.C. (1988) Adoption of financial management practices: A program assessment. *Journal of Extension* 36(2), 517–530.

Hanemann, W.M. (1984) Welfare evaluations in contingent valuation experiments with discrete responses. *American Journal of Agricultural Economics* 66, 332–341.

Kenya (2001a) *National Agriculture Extension Policy*. Ministry of Agriculture and Rural Development, Government Printers, Nairobi, Kenya.

Kenya (2001b) *Poverty Reduction Strategy Paper*, 2001–2004. Ministry of Planning, Government Printers, Nairobi, Kenya.

Kenya (2001c) *Kenya Population Census*, 1999. Government Printers, Nairobi, Kenya.

McFadden, D. (1994) Contingent valuation and socio-choice. *American Journal of Agricultural Economics* 79, 689–708.

Messonnier, M.L., Bergstrom, J.C., Cornwell, C.M., Teasly, R.J. and Cordell, H.K. (2000) Survey response-related biases incontingent valuation: concepts remedies and empirical application to valuing aquatic plant management. *American Journal of Agricultural Economics* 82(2), 245–486.

Millar, D. (1994) Experimenting farmers in northern Ghana. In: Scoones, 1. and Thompson, J. (eds) *Rural People's Knowledge, Agricultural Research and Extension Practice*. International Institute for Environment and Development, Intermediate Technology Publications, Southampton Row, London, pp. 103–105.

Smith, L.D. (1997) Decentralization and rural development: The role of public and private sector in the provision of agricultural services. *Technical Consultation on Decentralization*, FAO, Rome.

Thrupp, L.A. and Altieri, M. (2001) *Innovative Models of Technology Generation and Transfer: Lessons Learned from the South*. Kluwer Academic Publishers, Dordrecht, The Netherlands.

Wangila, J. and Swallow, B. (2001) Design principle for land and water shade in West Kenya. Discussion paper 2001. *International Center for Research and Agroforestry (ICRAF)*, Nairobi, Kenya.

# 6 Decentralization of Pastoral Resources Management and Its Effects on Environmental Degradation and Poverty: Experience from Northern Kenya

K. Munyao and C.B. Barrett

## 1 Introduction

Growing concerns about persistent poverty and environmental sustainability have helped fuel efforts at decentralizing governance throughout the developing world. The 1992 Earth Summit in Rio de Janeiro brought widespread calls for greater community participation and equity in natural resources management and sustainable development planning. These pressures have grown amid institutional reforms fostered by movements towards democratization and market-based economic policy, spurred by, among others, the Bretton Woods institutions (the International Monetary Fund and the World Bank) in the last two decades of the 20th century (Goumandakoye and Mathu, 2003). Ironically, however, in many cases decentralization has been used by national governments not as a means to cede authority to local subjects, but rather to extend control still deeper into local community life and resources management, while still reaping the political capital associated with the rhetoric of bringing government services and development closer to the people. Often this involves the subtle but real transfer of influence, even control, from customary users of the resources to newcomers with better connections to government representatives.

This raises a subtle but important distinction between 'decentralization' and 'deconcentration'. The former involves the delegation or devolution of authority from a central government to lower levels in a political-administrative and territorial hierarchy so as to allow for greater participation by local peoples and their direct representatives (Agrawal and Ribot, 1999). The latter refers to mere relocation of power to local branches of the central government that have minimal downward accountability as they remain primarily responsible to the central government (Ribot, 2002). The two processes may be in play at once, with decentralization of some authority to certain subpopulations going hand in hand with deconcentration with respect to other subpopulations. Naive enthusiasm for decentralization without an appreciation for the distinction between

 *Decentralization and the Social Economics of Development* (eds C.B. Barrett, A.G. Mude and J.M. Omiti)

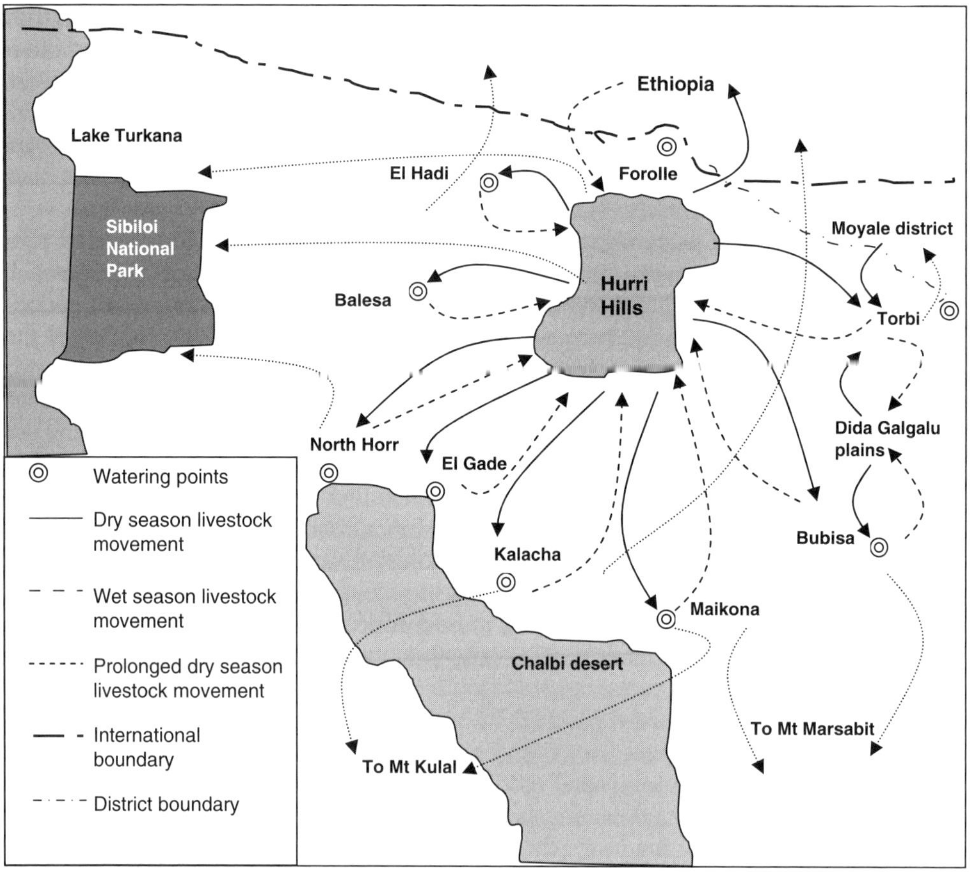


**Fig. 6.1.** Gabra land and seasonal livestock movements. (Adapted from the Global Environmental Facility – Indigenous Vegetation Project (GEF – IVP) Livestock migration routes map.)

sanctions spatio-temporal use of natural resources within Gabra territory. For example, destitute Gabra, Konso and Boran households who had lost livestock during drought were often allowed to settle temporarily in the Hurri Hills until they rebuilt their herds. But this was typically on a temporary basis and had to be negotiated with the traditional users of the Hills, the Gabra, through the *Yaa*.

Prior to 1974, the Hurri Hills had no permanent resident population as the nomadic Gabra pastoralists only utilized the rangelands during the wet season, descending to the lowlands during the dry season to access permanent water sources at springs and wells. With bimodal rainfall yielding two dry and two wet seasons of roughly 3 months each annually, this seasonal pattern of occupation preserved not only the rich pastures of the Hurri Hills' grazing lands, but also the Hills' biodiversity. The *Yaa* paid attention to indicators of biodiversity as it deliberated on resource use, with an eye towards the health of the ecosystem on which the community depended.

This system of land use and resources management continued until the 1980s, when a series of serious droughts combined with the initiation of the government's District Focus for Rural Development (DFRD) approach heralded central state involvement in local resources management issues (Bragdon, 1992). The droughts caused widespread herd loss, which forced many households to sedentarize, at least temporarily, as too small a herd size compromises spatial mobility in Gabra land, as in most of the East African ASAL (McPeak, 2005; Roth and Fratkin, 2005). Recently sedentarized pastoralists are typically the poorest in the area, as routinely manifested in wealth, nutritional and income data (Roth and Fratkin, 2005). But as civil service, NGO and small-scale commercial jobs emerge in more established settlements, the sedentarized become increasingly differentiated into the relatively economically successful, who choose remunerative non-pastoral livelihoods while typically still maintaining (even accumulating) herds, and the destitute, who are driven by misfortune to sedentarize (McPeak and Little, 2005). Thus, attending to the pressing needs of newly sedentarized, destitute herder households was a priority for government and humanitarian relief and development NGOs in response to a series of droughts, especially the 1984/85 drought in the Hurri Hills region. Addressing these humanitarian needs is made considerably easier when the target population is stationary and thus easier to reach. Hence, NGOs have historically encouraged sedentarization in order to facilitate services provision (and, among religious NGOs, evangelism). Once such settlements are established, opportunities often emerge for non-pastoral livelihood that subsequently complicates geographic targeting of the poor in the ASAL.

Central government's presence in the area grew markedly under the DFRD strategy, based on the district, divisional, location and sub-location administrative units. The District Commissioner (DC) sits at the apex of the district hierarchy within the Office of the President (OOP) as a political appointee and, under the strategy, has authority over sectoral field officers through his position as chair of the District Development Committee (DDC). Decisions concerning the prioritization and funding of development projects are made within the DDC. Thus, the DC wields significant power over district resources. As Bragdon (1992) notes, it is somewhat ironic that despite commitments to autonomy for the districts and participatory development, government policy states that the position of the DC should be filled by an officer from outside the district in which he or she serves. Chiefs and assistant chiefs are responsible for implementing policies and programmes at the location and sub-location levels, respectively. The Chief's Authority Act bestows significant authority upon the chief, who exercises control over field officers of sectoral ministries operating at the location level. Where, in the absence of environmental policy per se, the local environment is subject to the policies of sectoral ministries, the chief holds considerable power through his position of authority over local ministry field officers. In addition to administrative responsibilities, the chief also acts as a spokesperson for

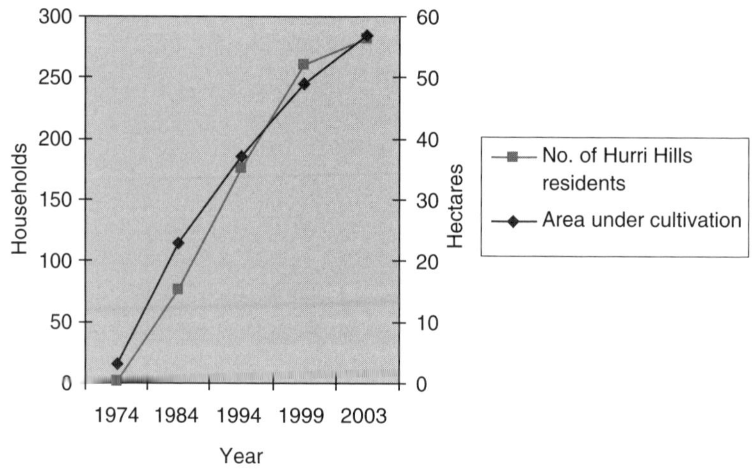


**Fig. 6.3.** Population and acreage under cultivation on Hurri Hills (1974–2003).

administrative areas have thus created contentious resources use boundaries that have complicated administrative tasks and fuelled tension and insecurity over pastoral resources use and land tenure between communities (Haro *et al*., 2005).

The creation of formal, local government structures has caused an overlap of the formal administrative and resources management institutions of government with the traditional authority of elders, such as exercised through the *Yaa*. Although government tolerance of the continued presence of traditional elders is a compromise aimed at gaining community support and building consensus on pastoral resources management, in reality the *Yaa*'s present influence over natural resources management has become minimal. Permanent residents' self-interest proves a major motivating factor in the deliberations of the formal institutions of local government since the sanctioning authority is concentrated in the hands of the administrative chief, a central government appointee.

The creation of entities such as land allocation committees has occurred within a wider framework in which more and more community-based groups are emerging. For example, the number of community groups officially registered in the Hurri Hills has increased from virtually none in 1975 to 25 in 2004 (Fig. 6.4). On the positive side, some of these groups provide a voice for otherwise voiceless groups, especially women and young people, in highly paternalistic pastoral communities. But they also created alternative centres of power within the community, liable to manipulation by the local settled elite, thereby undermining the effectiveness of existing community institutions such as the *Yaa*.

The emergence of a better-educated, more politically aware class of elites among settled local communities, on one hand, and the weakening of traditional institutions prevalent among non-resident, nomadic populations, on the other, have been a major source of conflict and ambiguity in community-based resources management efforts in the Hurri Hills, as in other parts of

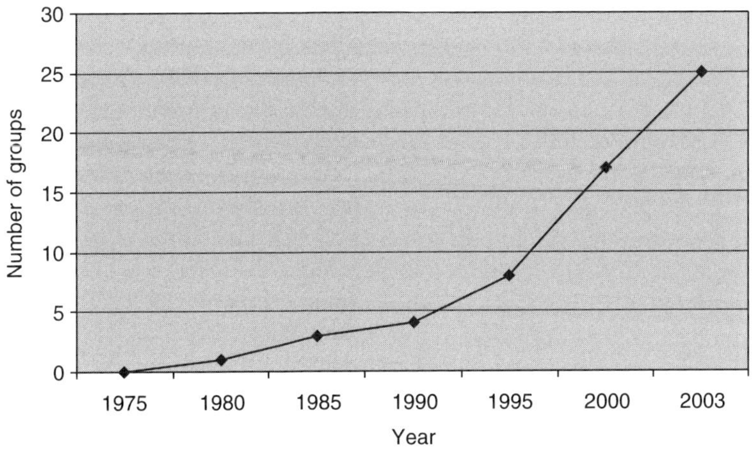


**Fig. 6.4.** Registered community groups in Hurri Hills (1975–2003).

northern Kenya and the broader East African ASAL. According to Yirbecho *et al.* (2004, p. 1):

> [R]esource conflict in this region appears to be associated more with the rise of nontraditional land uses, especially crop cultivation in [areas] traditionally used for grazing and watering herds, than with any growth in herd sizes associated with livestock cycles or growing pastoralist populations. Traditional pastoral communities tend to have fewer resource-related conflicts than communities experiencing a rise in crop cultivation. The traditional pastoral system appears more capable of mitigating resource-related conflicts and of resolving them when they do occur, while such conflicts appear to be more frequent and less easily resolved where land use patterns are shifting away from traditional extensive grazing systems towards more diverse land use systems incorporating cultivation as well as grazing.

The problem arises because most initiatives in decentralization and community empowerment driven by central government and external donors (i.e. from the top down) treat cultivators and herders as if they were a single entity, disregarding their varied sources of legitimacy and livelihood as well as the inherent conflict in their ambitions for the natural resources base on which they each depend. In such cases, enforcement of community sanctions fails to achieve significantly because the two groups differ in terms of cultural norms and legal status, despite overlaps in resources management mandates. Most of the emerging community groups arise from empowerment and development initiatives spearheaded by development agencies working with government and thus are prone to (often inadvertent) capture by the settled population, and at times are at variance with the existing traditional institutions.

Consider the example of the Environmental Management Committee (EMC) intended to spearhead environmental management and biodiversity conservation efforts. The EMC concept, introduced in the area by the GEF project in 2003, drew inspiration from a German donor agency's (GTZ) experiences working with the pastoral Rendille community in south-west Marsabit district.

Through the Marsabit Development Program (MDP), GTZ spearheaded the establishment of EMCs to deal with problems of natural resources management and localized environmental degradation caused by overstocking and settlement at water points and trading centres (Haro *et al.*, 2005). Ideally, the EMC membership comprises traditional leaders, women and youths within an identifiable neighbourhood based on shared resources. The EMC is charged with mobilizing and raising environmental awareness for user groups by helping organize meetings to elaborate and disseminate environmental management protocols aimed at minimizing natural resources–related conflicts and facilitating participatory assessment of implemented actions and measures (Haro *et al.*, 2005). In most other parts of Marsabit district where this model is used, the EMC mainly deals with pastoralists and settled communities near water points and trading centres that comprise only one ethnic group.

In the case of the Hurri Hills, however, the presence of ethnically diverse resident cultivators (Boran and Konso) and non-resident pastoralists (Gabra) greatly complicates the EMC's work and undermines its legitimacy. Since the EMC is exclusively made up of year-round Hurri Hills residents, its composition is mainly Boran, while the traditionally recognized natural resources decision makers on the Hills had been the nomadic Gabra elders. This has posed a major challenge to the effectiveness of the EMC due to its perceived lack of legitimacy among the nomadic Gabra pastoralists. The limited cultural precedence for a body such as the EMC to define new rules for resources use has led some community members to refuse to accept the final authority of the EMC (Haro *et al.*, 2005).

The EMC's dependency on elders or the authority of a local chief to enforce sanctions for non-compliance with conservation by-laws has also been a problem. In addition, its inability to offer incentives for compliance has seriously compromised its effectiveness. Enforcement of resources use restrictions traditionally relied on community elders who induced compliance out of fear of being ostracized or cursed. Where traditional institutions have been undermined and support of the elders is not assured, the effectiveness of community-based efforts has become doubtful because such social sanction mechanisms become less effective.

## 3 Conclusions and Recommendations

Decentralization of land use authority in the Hurri Hills seems to have had several unintended and undesired consequences. Rather than devolving authority to established, traditional institutions with credibility among extant resources users, the central government followed a pattern of deconcentration, extending authority to spatially dispersed agents of the central government through the creation of new, more disaggregated jurisdictions, locations and sub-locations. These were staffed typically by outsiders but brought the full powers of the state to the local level.

This new, local-level central government authority both reflected and fostered increased settlement and cultivation in the Hurri Hills, investing newly settled households and non-traditional livelihood with unprecedented influence.

This has, however, impeded traditional, non-resident pastoralists' access to customary livestock migration routes and displaced them from critical wet season grazing areas, thereby increasing their vulnerability to drought. This transition has helped fuel increased conflict over water resources and crop damage by livestock. Increased settlement and sedentarization has accelerated localized environmental degradation due to greater demand for fuel wood and tim ber for building, as well as increased soil erosion due to cultivation (Munyao, 2005). Although no firm data exist by which one can conclusively establish the effects of sedentarization and deconcentration/decentralization of government authority on either the quality of the natural resources base or on poverty and well-being in the area, widespread local perception, including that of many NGOs working in the region, and the launch of the World Bank GEF project on the Hurri Hills suggest that deconcentration of government authority has been associated with deterioration in both socio-economic and environmental indicators over the last 30 years.

The key to effective decentralization to empower the poor is increased, broad-based participation in local decision making concerning common pool resources and public goods and services provision. Downwardly accountable or representative authorities with meaningful discretionary powers are the basic institutional elements of decentralization that lead to efficient, equitable, sustainable and credible resources management (Agrawal and Ribot, 1999). Effective decentralization in the area of natural resources management requires these same elements. However, case studies from around the world indicate that the institutional arrangements necessary to bring about effective decentralization (in terms of both improvement of poor peoples' livelihood and conservation of critical natural resources on which they depend) are often not present in so-called decentralization reforms, too many of which turn into exercises in deconcentration alone (Agrawal and Ribot, 1999; Barrett *et al.*, 2001; Ribot, 2002, 2004; Goumandakoye and Mathu, 2003). In addition, communities' capacity to self-govern natural resources in a way that promotes conservation and equity cannot always be assumed (Barrett *et al.*, 2001). Community-based methods work best when there are strong local systems of social control to enforce access restrictions, as rules enforcement appears to be the core element of effective management of common pool resources (Gibson *et al.*, 2005). Unfortunately, some government-led decentralization processes have undermined, rather than reinforced, pre-existing local systems of social control by imposing alternative governance systems and effectively transferring authority to new stakeholders whose livelihood systems are often less well adapted to local natural resources conditions.

In the case of the Hurri Hills of northern Kenya, deconcentration appears to have created new pressures for sedentarization and resources overuse – manifest in higher human and livestock populations – as well as more cultivation and localized resources degradation in the Hills, and to have displaced traditional Gabra authorities and disfavoured non-resident Gabra pastoralists in favour of permanent Boran settlers.

Cases such as this one raise the question of whether the national constitutional review process might be used to advocate for proper devolution of state authority to local peoples. It also underscores how, at least in pastoralist regions, decentralization of governance systems needs to be considered within the overall framework of a mobile community dependent on the spatio-temporal allocation of natural resources with loosely defined boundaries. Without taking this into consideration, groups of pastoralists will likely continue to be sidelined in the decentralization process, leading to inadvertent deconcentration and the eventual dispossession of their natural resources.

During decentralization, the government and development agencies also need to pay closer attention to the array of local interests and the prospect for competing centres of power within local jurisdictions and what this might mean for land use patterns, equity and security. Indeed, support for decentralized resources management may require a reconceptualization of the role of the state and other local-level institutions in natural resources management.

In many cases, self-organized institutions may do better than those imposed by the state (Finke, 2000). The persistence of traditional institutions, such as the Gabra *Yaa*, despite years of government-sanctioned competition, attests to their resilience. Their overarching influence on the sociopolitical and economic life of the Gabra community locally and nationally reinforces their credibility. It may thus prove more effective to link political and development decentralization strategies to existing community institutions, not only to strengthen them, but also to enhance the legitimacy and viability of such strategies. In some cases, state intervention may be necessary to address the needs of otherwise marginalized groups in the community. But decentralization needs to be backed by enforceable sanction and reward mechanisms for non-compliance and compliance, respectively, and by a similar decentralization of the financial resources necessary to implement those mechanisms.

Perhaps the clearest lesson from the Hurri Hills example is the need for greater institutional support and capacity development within existing pastoralist institutions in order for them to be able to undertake policy advocacy. The highly disenfranchised nature of many pastoral communities and the state's deeply ingrained bias against pastoral communities and regions militate against their ability to positively advocate for policy change, since influencing the policy environment is a complex task and is highly dependent on the willingness of governments to listen to its citizens (IIED, 2003). Ultimately, the success of decentralization and effective management of pastoral resources will depend to a great extent on appropriate policies, the capacity of local institutions as well as the political will to undertake decentralization in a manner that truly empowers the traditionally voiceless and relatively poor communities of pastoralists.

## Acknowledgements

For financial support that made this work possible, we thank the Government of Japan, the World Bank, the Mario Einaudi Center for International Studies, the Cornell Institute for International Food, Agriculture and Development, the Cornell Institute for Policy Analysis, and the United States Agency for International

Development, through the Pastoral Risk Management (PARIMA) project of the Global Livestock Collaborative Research Support Program (GL CRSP), funded by the Office of Agriculture and Food Security, Global Bureau, USAID, under grant number DAN-1328-G-00-0046-00, and the Strategies and Analysis for Growth and Access (SAGA) cooperative agreement, grant number HFM-A-00-01-00132-00. We also thank Chuck Geisler, John McPeak, Andrew Mude and John Omiti for comments on earlier drafts. Our greatest thanks go to the Kalacha and Hurri Hills communities and to the dedicated organizations that work in those locations. Their assistance made this work feasible and meaningful. Any remaining errors are the authors' alone.

## References

Agrawal, A. and Ribot, J.C. (1999) Accountability in decentralization: a framework with South Asian and African cases. *Journal of Developing Areas* 33, 473–502.

Barrett, C.B., Brandon, K., Gibson, C. and Gjertsen, H. (2001) Conserving biodiversity amid weak institutions. *BioScience* 51(6), 497–501.

Bragdon, S.H. (1992) Kenya's legal and institutional structure for environmental protection and natural resource management: an analysis and agenda for the future. McNamara Fellowships Program Paper. Economic Development Institute, World Bank, Washington, DC.

District Development Officer (1997) Marsabit District Development Plan 1997–2002. Ministry of Planning and Development, Kenya.

Finke, P. (2000) *From 'Common Property' to Open Access: Changing Pastoral Land Tenure in Post-Socialist Mongolia.* Eighth Conference of the International Association for the Study of Common Property, Bloomington, Indiana.

Gibson, C.C., Williams, J.T. and Ostrom, E. (2005) Local enforcement and better forests, *World Development* 33(2), 273–284.

Goumandakoye, M. and Mathu, W. (2003) *Decentralized Governance of Natural Resources.* UNDP Dryland Development Centre: 33.

Haro, G.O., McPeak, J.G. and Doyo, J.G. (2005) Linkages between community, environment and conflict management: experiences from Northern Kenya. *World Development* 33(2), 285–299.

IIED (2003) *Shared Management of Common Property Resources in the Sahel: A Regional Action-Research Programme.* International Institute of Environment and Development: 24.

Lane, C.R. (1998) *Pastoral Land Tenure in East and West Africa.* UNRISD-IIED, Earthscan, London.

McPeak, J.G. (2005) Individual rationality and collective rationality in pastoral production: evidence from Northern Kenya. *Human Ecology* 33(2), 171–197.

McPeak, J.G. and Little, P.D. (2005) Cursed if you do, cursed if you don't: The contradictory processes of pastoral sedentarization in Northern Kenya. In: Fratkin, E. and Ruth, E. (eds) *As Pastoralists Settle: Social, Economic, and Health Consequences of Pastoral Sedentarization in Northern Kenya.* Kluwer, Dordrecht, The Netherlands.

Munyao, K. (2005) Pastoral land use changes in northern Kenya: challenges and opportunities: land use change in the Hurri Hills and the implications for pastoralism and biodiversity conservation in Marsabit district, Kenya. Master's thesis, Cornell University, Ithaca. New York.

Ribot, J.C. (2002) *Democratic Decentralization of Natural Resources: Institutionalizing Popular Participation.* World Resources Institute, Washington, DC.

Ribot, J.C. (2004) *Waiting for Democracy: The Politics of Choice in Natural Resource Decentralization.* World Resources Institute, Washington, DC.

Robinson, P.W. (1985) Gabra nomadic pastoralism in the nineteenth and twentieth century northern Kenya; strategies for survival in a marginal environment. Northwestern University: 21–36.

Roth, E. and Fratkin, E. (eds) (2005) *As Pastoralists Settle: Social, Economic, and Health Consequences of Pastoral Sedentarization in Northern Kenya.* Kluwer Academic Publishers, Amsterdam, The Netherlands.

Schwarz, H.J. (1991) Marsabit district, Kenya range units. Nairobi, Ministry of Livestock Development: map set for the Range Management Handbook of Kenya; Marsabit District.

Sobania, N.W. (1979) Background History of the Mt. Kulal Region of Kenya. IPAL Technical Report. United Nations Education and Scientific Organization (UNESCO), Nairobi: 43–46.

Tablino, P. (1999) *The Gabra: Camel Nomads of Northern Kenya.* Paulines Publications Africa, Limuru, Kenya.

Yirbecho, A.T., Barrett, C.B. and Gebru, G. (2004) Resource conflict in the rangelands: evidence from northern Kenya and southern Ethiopia. USAID Global Livestock CRSP.

# 7 Linkages between Community, Environment and Conflict Management: Experiences from Northern Kenya

G.O. HARO, G.J. DOYO AND J.G. MCPEAK

## 1 Introduction

There has recently been a great deal of attention paid in the literature to the issue of local participation in natural resource management in Africa (Ingles *et al*., 1999; Turner, 1999a; Kellert *et al*., 2000; Moore *et al*., 2000; Barrett *et al*., 2001; Environment and Natural Resources Team, 2002; Ribot, 2002). Various studies have illustrated that community participation is a critical component of efforts that attempt to cause positive economic and ecological change in African communities. This study contributes to the growing literature on community management of natural resources by presenting information on such a programme in a pastoral area of northern Kenya. It illustrates how local participation led the natural resource management project to take an unexpected route to achieving positive economic and ecological change by encompassing issues of conflict management.

This study also contributes to the literature on common property management in risky production environments. As is increasingly understood, the finding that common property management regimes function best with clearly defined boundaries and membership (Ostrom, 1990, 1992) is in conflict with the finding that such clear definitions can be welfare-reducing in highly variable environments (van den Brink *et al*., 1995; Vedeld, 1998; Nugent and Sanchez, 1999; Goodhue and McCarthy, 2000). This has led to a recent, growing interest in pastoral development efforts that strengthen management structures while still providing for flexibility in land use patterns (Niamir-Fuller and Turner, 1999; Turner, 1999, 1999b; Fernandez-Gimenez, 2002). This study identifies some of the promise and notes some of the challenges of conducting such an effort to build land use management plans on existing social structures.

The findings described in this chapter also contribute to a growing literature on the relationship between environmental variables and conflict. It is recognized in the literature that natural resource management and conflict management are closely related (FAO, 2000, 2001; Lind and Sheikh, 2001; Lind, 2002; Castro and Nielsen, 2003). The literature to date has largely focused on how environmental scarcity leads to increased conflict and how natural resource management plans can be designed to manage conflict (Homer-Dixon, 1991, 1994; Lind and Sturman, 2002). The data provided by this study give a different perspective on the relationship between environmental variables and conflict as the study illustrates how conflict management can be a precondition for implementing a resource management plan. This approach reflects some of the findings in the recent literature on development efforts in insecure pastoral areas. It is increasingly recognized that addressing insecurity is a critical first step for any development efforts designed to improve pastoral welfare in such areas (Kratli and Swift, 1999; Kenya Human Rights Commission, 2000; Odhiambo, 2000; Galaty, 2002; Lind, 2002). As we will illustrate below, what began as a programme to improve the well-being of pastoral populations through improving resource management evolved to become a programme that focused on reducing insecurity, thus both enhancing well-being and allowing the potential for improved environmental management.

An important element of the case study we present is that adoption of a community-driven approach led the implementing agency to confront issues of conflict management that it had not anticipated in the original programme design. The study illustrates that flexibility and adaptability are not only relevant to understanding the behaviour of pastoralists but also critical to designing effective participatory approaches for community natural resource management.

Section 2 briefly describes the study area. Section 3 places community management of natural resources by pastoral populations in a historical context. Section 4 portrays the management structure of natural resources in the study area, with emphasis on ambiguities arising over geographic boundaries. Section 5 discusses insecurity in the study area. Section 6 focuses on environmental management efforts in Marsabit district, as well as on the evolution of a German donor agency (GTZ)-funded project in the area. Section 7 closes the chapter with a discussion of the prospects for the future with this effort, and also summarizes the larger themes of policy relevance illustrated by the case study.

## 2 The Study Area

Marsabit district is in the Eastern Province of Kenya. It borders Ethiopia and Moyale district to the north, Lake Turkana and Turkana district to the west, Samburu district to the south and Wajir and Isiolo districts to the east. The district is the second largest in the country after Turkana. The estimated population is 125,000. Approximately 75% of the district is classified as rangelands and the main mode of land use is extensive grazing. The district is a home to a number of ethnic groups like Boran, Gabra, Rendille, Samburu, Ariaal, Turkana, Burji and Dassenetch. Alliances and hostilities vary from community to community and change over time.

The pastoral groups considered in this chapter live in the arid and semiarid areas of this district and are interrelated in a variety of ways. Rendille and Gabra tend to specialize in camel, goat and sheep pastoralism, and their livestock are highly mobile. Boran, Samburu and Ariaal focus more on cattle production in higher rainfall areas, and are less mobile than camel-based pastoralists. Rendille, Gabra and Boran are Cushitic languages and Samburu and Turkana are Nilotic languages. Rendille and Gabra share cultural practices and clan histories. Rendille and Samburu are linked by a history of cooperation, the outgrowth of which is seen in the Ariaal group who combine elements of both Rendille and Samburu culture (Spencer, 1973; Fratkin, 1991).

All groups in Marsabit have faced severe challenges in the last 30 years. Beginning around 1970, there has been an increase of population in permanent settlements in this district around water points. Households that were nomadic prior to 1970 have settled for a variety of reasons. One is the loss of animals in droughts, which were experienced in 1969–1973, 1980 and 1984, leading to household herd sizes that were insufficient for maintaining mobility. A second reason is the increased provision of public services in towns, such as health centres, schools and food aid, as well as the increased economic opportunities offered by towns. Finally, insecurity has led to settlement and concentration of grazing in areas around towns as it is no longer possible to use regions that were formerly grazing areas given the threat of armed raids. People settled in and around towns to provide mutual security.

## 3 Community Management of Natural Resources in Pastoral Areas

The ability of pastoral populations in east Africa to manage their own resources has long been viewed with scepticism. In large part, this scepticism results from the view that pastoral production is the cause of degradation and desertification, due to the inherent incentive problems of common property production and the cultural values of pastoralists (Pratt and Gwynne, 1977; Doran *et al.*, 1979; Jarvis, 1980). Due to the nature of the production system and the cultural context of production, it has been proposed that herders will accumulate more animals than is optimal from an environmental perspective.

Policies influenced by this view were common in the colonial and early post-independence period. Sobania (1979) quotes colonial era documents from the 1930s arguing that since pastoralists in northern Kenya own far too many animals from an environmental point of view, veterinary programmes will be counterproductive. Rather, the document suggests that 'a bit of disease now and then is to be encouraged in their stock provided it doesn't reach epidemic form' (Sobania, 1979, p. 180). Lipscomb (1955) summarizes the problems of the pastoral livestock sector of Kenya in one word – overstocking – and describes controlled grazing schemes to address this problem. Brown (1971) suggests that the objective of conservation can be met by combining destocking with the partial removal of human populations from semiarid areas to ease population pressure and with change in the diets of those who remain behind.

As the nature of pastoral production is posited as the underlying cause of degradation, this perspective holds out little hope that the pastoral population will be capable of addressing rangeland degradation. Barnes (1979, p. 51) argues: 'The future of large tracts of Africa thus depends, in the first instance, on drastic changes in traditional attitudes towards land-use among relatively unsophisticated and uneducated indigenous peoples. This can only be brought about by concerted and well-planned programs of rural reform and education.' Walker (1979) supports the view that such programmes will require direction from outside the pastoral sector. He argues that since people with initiative and high capabilities are attracted away from semiarid regions to higher potential zones, 'semi-arid ecosystems have, therefore, often been managed by a segment of the population which constitutes the least capable, least innovative group, often disinterested in what they are doing, but not capable of changing their circumstances' (Walker, 1979, p. 3).

These views influenced the design of development programmes in pastoral areas. In a review of World Bank pastoral development schemes, de Haan (1994) describes the development efforts arising from this approach as falling into the 'Ranching Phase'. This phase began in the colonial period and lasted until the mid-1980s. It involved the transfer of western technology to arid African rangelands, and involved a high degree of capital investment and direction by expatriate staff (Moris, 1998). The objective was to transform pastoral production into commercialized ranching, which it was believed would simultaneously increase human welfare by commercializing livestock production systems and reverse environmental degradation by addressing common property incentive problems.

The outcome of these efforts was disappointing (Sandford, 1983; World Bank, 1985). Scoones (1995, p. 3) describes the experience of development efforts in pastoral areas as one of 'unremitting failure . . . millions of dollars have been spent with few obvious returns and not a little damage'. Baxter (cited in Brandstrom, 1985, p. 41) states that he pleaded in vain 'for someone to cite just *one* pastoral development project which had been even partially successful, so that we might learn from success if we refused to do so from failure'. One of the main lessons learned from this failure, albeit slowly, was that the projects implemented during this era lacked support of the affected population (Lusigi, 1981; de Haan, 1994). Marty writes: 'The vast investment poured into the livestock sector has failed to achieve anything, because of the exclusively technical definition of the activities and the indifferent participation of the producers' (quoted in Sylla, 1995, p. 135).

Frustration with the failure of development efforts led to a growing appreciation of the need to involve pastoral organizations in programme design (de Haan, 1994; Sylla, 1995). In the first phase of this effort, pastoral organizations were largely viewed as institutions through which a project message would be disseminated. This extension-oriented approach was designed as a means of delivering information to producers via these organizations (Butcher, 1994).

These efforts have been modified over the last 10–15 years to place increased emphasis on natural resource management by pastoral organizations (de Haan, 1994; Pratt *et al.*, 1997). In part, this reflects the overall move in the field of development to participatory methods (Chambers, 1997). Growing emphasis has been placed on understanding traditional environmental management practices (Oba, 1992). Increased appreciation for traditional management practices often developed when pastoral extension officers found that their work among pastoral populations led to an increased appreciation of traditional pastoral practices on the part of the extension agent, rather than the adoption of the extension message by the pastoralists (Akabwai, 1992).

## 4 Land Use Management, Mobility and Boundaries in the Study Area

While the various ethnic groups differ in the details of their social organization, cross-cultural comparison suggests that are common themes to the pattern of decision-making authority over natural resources within pastoral communities. All of the groups in this area have some form of a nested decision-making structure that is not characterized by a clear hierarchy of authority. The smallest decision-making group is the household. Households make decisions over managing their labour force, specific grazing route, and house and livestock enclosure maintenance.

The next level of decision making takes place at the camp level. A collection of households settled in the same immediate area make decisions as a camp related to issues such as managing local drinking water sources, watering order at water points, maintenance of water points, direction in which animals will be taken to graze, and defence against human and wildlife predation.

Representatives from different camps using a given area are sometimes called together to make decisions at a level that can be thought of as neighbourhood associations. Neighbourhood groups deal with many of the same issues as camp-level decisions, but focus on coordinating and managing the efforts of multiple camps using many of the same resources.

Finally, community decision making also takes place in meetings where neighbourhood representatives discuss issues related to the grazing area used by residents of different neighbourhoods. As households and livestock are mobile, the grazing area can be thought of as the area within which a household can potentially move their household and animals. Such meetings discuss long-range migration strategies, management of dry season reserve areas and issues of conflict with herders from other grazing areas, be they of the same tribe or of other tribes.

These different decision-making structures defined at differing levels of social organization within a community may have overlapping rights to make decisions impacting a single resource. For example, a single water pan or grove of trees can have household, camp, neighbourhood and grazing area decision makers draw on cultural precedent to support their right to make decisions over how the resource should be managed. This presents a challenge to community management of environmental resources, as it is not always clear which particular level of decision making is the most effective or appropriate for addressing management issues of a particular resource.

This challenge is further exacerbated by the fact that for a given decision-making structure defined by the level of social organization, multiple groups at this level, both within and across ethnic groups, may claim rights to a single resource. Strict definitions of geographical boundaries are not emphasized by pastoral groups. The concept of a 'boundary' does not have a simple analogue for the pastoral cultures of northern Kenya (Schlee, 1990). When describing the land associated with a particular group, the discussion focuses on a specific location associated with the group (physical camp, water point, geographical formation) rather than an area with clearly delineated boundaries.

Oba (1992) and Robinson (1985) describe how this ambiguity over geographic boundaries at the ethnic level is present within ethnic groups. Subgroups within Rendille and Gabra have specific geographical migration routes that they tend to follow. However, a particular group's association with a route does 'not preclude movement of other clans into the same grazing area' (Oba, 1992, p. 42). Schlee describes the ambiguity in boundaries across ethnic groups. When he asked a Rendille elder what constituted 'Rendilleland', the elder described a set of places where Rendille can currently be found. When asked explicitly what constituted the boundaries of 'Rendilleland' in northern Kenya, the elder responded that the only real boundary was 'one of fear' (Schlee, 1989, p. 24); where one had gotten too close to hostile neighbours. When pressed to compare the concept of administrative boundaries (Marsabit district as compared to Samburu district) with the concept of boundaries between Rendilleland and the land belonging to the neighbouring Samburu, the elder responded that these were different concepts. He described Rendilleland and Samburuland as 'inside each other . . . they are mixed up' (Schlee, 1989, p. 24). The 'land of the Rendille' was a separate concept in his mind from the clearly defined boundary separating administrative districts which was viewed as a concept introduced during the colonial era and carried through to the present day.

Whether the issue is multiple claims to the wood from a fallen tree across camps, multiple claims on a water point across neighbourhoods or multiple claims to a dry season reserve area across grazing areas, overlapping management authority, both within socially defined levels of organization and across spatially defined areas, present a challenge to community management of the environment. We now turn to one prominent aspect of this challenge that occurs when there are unclear boundaries across ethnic frontiers.

## 5 Insecurity

A major issue to be confronted when working in pastoral areas of east Africa is insecurity. Galaty (2002) finds that addressing insecurity in northern Kenya is critical due to the impact of escalating local cycles of conflict. Kratli and Swift (1999) discuss alternative theories about the source of this violence, noting a gradual erosion of elders' authority, failure of the state to provide security, proliferation of small arms and greater integration into the national political and economic sphere. While conflict between ethnic groups in this area has always been present (Sobania, 1979; Robinson, 1985), there has been a qualitative transformation from battles among spear-wielding warriors into indiscriminate assaults on populations using semi-automatic weapons over the last 30 years

(Kratli and Swift, 1999; Kenya Human Rights Commission, 2000; Lind and Sheikh, 2001, Galaty, 2002). This change in the nature of conflict in Marsabit has contributed to a climate of fear and insecurity in the region, and left a legacy of hostility and mutual suspicion.

Lind (2002, p. 1) notes that overall 'there is an expanding recognition that peace and security are fundamental to Africa's social and economic renewal'. This is increasingly being realized by active donors in pastoral areas. Odhiambo (2000) describes the experience of a development project trying to work in an insecure pastoral area of Uganda. The programme began by trying to address other development needs in an insecure area, but was eventually led to the realization that to meet these other needs they had to directly address the issue of insecurity.

Confronting the issue of insecurity is particularly important for programmes attempting to address environmental issues. As is increasingly recognized, rangelands in northern Kenya are characterized by localized rather than widespread overstocking. Total rangeland resources are more than adequate to support the aggregate livestock herd, but the majority of the rangeland is either underused or completely unused due to insecurity. Environmental change occurs due to the poor spatial distribution of animals rather than the absolute number of animals owned by the herders (O'Leary, 1987; Schwartz *et al.*, 1991; Milimo *et al.*, 2002; McPeak, 2003).

Community management of natural resources thus takes place in an environment where the boundary of the resource area is often defined as 'one of fear'. This presents both an opportunity and a challenge to such management structures. The opportunity arises from the fact that positive environmental change can be brought about by reducing the level of fear so that grazing pressure is spread to areas currently underused due to insecurity, thus allowing currently overused areas to recover. The challenge is that communities are often not able to stop the cycle of violence that they find themselves in, and may need outside facilitation. We describe a case study of how environmental management programmes in northern Kenya faced these opportunities and challenges.

## 6 Environmental Management in Marsabit District

### 6.1 The integrated resource assessment and management plan

The Integrated Project for Arid Lands (IPAL) was a UNESCO-funded project operating in Marsabit, Kenya, from 1976 to 1986. The project was established with a focus on 'the arid lands of Kenya both for the support of their indigenous people and in the economy of the country as a whole, and because these lands were gravely threatened by desertification through misuse' (Lusigi, 1981, p. 7). The objective of the project was to develop a series of management steps that could be demonstrated and extended to the pastoral population through training and education. A variety of studies were produced during the life of this project covering a broad spectrum of topics such as the soils, hydrology, vegetation, livestock, history, and social and economic organization of the study area.

The culmination of the project was an integrated resource assessment and management plan produced in 1984. The management plan (Lusigi, 1984) describes in detail how the natural resources of the south-western portion of Marsabit should be managed in light of the scientific evidence gathered over the life of the project. The rangelands of the project area were divided up into distinct rangeland units based on vegetation, and differing management plans for each rangeland unit were defined. The plan explicitly stated that addressing the priorities of the pastoral population was critical for project success. A list of priority interventions expressed by the community was noted in the plan, with water development, market development, improved health care, veterinary service, improved security, improved leadership and drought assistance identified as the most important issues to be addressed. However, the plan says that while these are important 'some items of obvious importance are absent from the people's list primarily because they have no experience of their value. Such are the need for grazing control, means of storing wealth other than "on the hoof" (i.e. banking facilities, and the registering of tribal rangelands in order to put them on a firm legal basis)' (Lusigi, 1984, p. 486).

Recognizing that 'the success of the plan depends on the attitude [the pastoralist people] adopt towards the whole plan' (Lusigi, 1984, p. 499), the plan describes an extension component with the following objectives:

**1.** To maintain direct contact with the Rendille pastoralists;
**2.** To help the pastoralists to develop an understanding of the management programmes, and thus predispose them to effective cooperation;
**3.** To teach them the need to conserve and rehabilitate their environment;
**4.** To gather and evaluate educational programmes (Lusigi, 1984, p. 617).

Outside of a few small-scale efforts to implement plan components, the management plan was never enacted. To some extent, this resulted from a series of struggles between project personnel and local politicians both due to debate about the distribution of benefits of the programme and the resources of this area and trepidation on the part of the political elite about the implications of raising community awareness. In addition, donors were hesitant to fund a wide-ranging management plan that simultaneously embraced livestock, pasture, human health, forestry, water, agriculture, education, security, leadership development, alternative income and marketing components. But perhaps the most important reason for the limited impact of the plan was the lack of integration of the target population in designing a management plan for their own future. The plan was premised on the view that the local population could be given proper incentives and education that would allow them to live in harmony with their environment. The design largely reflected the research findings of the project staff and little importance was accorded to the local population in defining their own future. While there was some effort to incorporate the goals and aspirations of the community, the extension programme described above largely viewed the population as a group that needed to be convinced of the validity of a scientifically derived plan.

That the pastoralists would have to have explained to them items of obvious importance they had not previously considered such as grazing controls was a particularly ambitious objective for an extension programme. Oba (1985) reports that only 16% of respondents from his survey of 167 elders and herders in the study area believed it possible for humans to cause soil degradation. Similarly, only 16% believed droughts were related to overgrazing. There was a major disconnect between the perceptions of the population and the view of the designers of a management plan that was to protect lands 'gravely threatened by desertification through misuse'.

## 6.2 Intra-ethnic environmental management committees

The German donor agency GTZ designed a successor to IPAL called the Marsabit Integrated Development Programme (MIDP) in 1990. MIDP (later shortened to MDP) defined four main areas of intervention: livestock production and marketing; natural resources management; human resources development; and farming systems development. In this study, we focus on the natural resources management component.

Within the natural resources component the main areas of focus were: provision of appropriate water sources in the underutilized grazing areas; demonstrating and supporting rangeland rehabilitation; and promoting application of useful traditional grazing practices. With regard to this final objective, MDP commissioned a study to identify and assess traditional grazing systems and to elaborate an extension and education plan for the pastoral population (Oba, 1992).

Following completion of this study, MDP embarked on promotion of useful grazing practices targeting local administrative and civic leaders, traditional leaders and primary schoolteachers at the neighbourhood level between 1993 and 1994. The hope was that the prominent leaders who participated in workshops would disseminate the importance of the practices through traditional channels of communication. This would lead neighbourhoods to define caretaker committees charged with the responsibility of day-to-day management of environmental matters within their neighbourhoods.

The approach of targeting key leaders who would then disseminate useful practices proved unable to respond to intricacies of community decision-making authority discussed previously. Local residents were expected to define action plans to implement environmental management programmes at neighbourhood levels. However, neighbourhood leaders found it extremely difficult to implement these plans. One reason was that within a community, there was no widespread acceptance of the legitimacy of the committee since there was no cultural precedent for such rule-making and enforcement. Further difficulties arose when bylaws initiated in a specific neighbourhood were not respected by pastoralists from other neighbourhoods. Pastoralists originating from other areas did not accept the legitimacy of rules defined for a given neighbourhood, as there was no traditional precedent for a neighbourhood to have exclusive claim to the resources in their area. Given the multiple layers at which individual herders felt they had rights to a particular area, ambiguity about what were the spatial boundaries of a given area and questions about whether local residents had a legitimate right to make decisions over natural resources, the efforts of neighbourhood leaders to impose resource management regimes began to appear more of a spark to resource use conflict than a measure to address land degradation.

MDP reassessed its effort to design neighbourhood-level natural resource management plans in 1995 due to these problems. A review of the programme was initiated to identify the underlying causes of natural resource degradation, deliberate on how to improve existing local-level resource management structures and initiate a consultative process on community-driven sustainable use of common resources through consensus building. Participants in these discussions were drawn from two administrative divisions covering 11 adjacent neighbourhoods in south-west Marsabit district.

The results of this review built on previous work at the neighbourhood level, but also addressed the growing issue of resource use conflict across neighbourhoods that had resulted from attempting to implement neighbourhood-level management plans. The main recommendations were the following: establish and support specifically designed Environmental Management Committees (EMCs) composed of elders, traditional leaders, women and youth within all identifiable neighbourhoods; mobilize and raise environmental awareness for user communities in all neighbourhoods; support workshops to elaborate and disseminate environmental management protocol within their neighbourhoods; initiate and support interneighbourhood discussions aimed at minimizing natural resources–related conflict between user groups; and facilitate participatory assessment of all implemented measures.

Between 1996 and 1998, MDP started implementing these recommendations. The project area was zoned into management units corresponding to traditional definitions of neighbourhoods. Community environmental awareness field days were conducted in each neighbourhood using posters depicting time-series environmental changes and degradation of known areas around them. The objective of conducting these field days was to analyze with neighbourhood residents specific changes in land use patterns; identify the causes and effects of such changes; identify the course of action to take to address these changes; and identify individuals from each camp who would form the EMC charged with managing the resources of the neighbourhood and would be viewed by residents as legitimate authorities.

By 1998 a total of 29 neighbourhood-based EMCs had been formed. A total of 10,150 households (out of 21,602 households in the whole area) were in some way involved in the formation of these management committees. The established EMCs had a total membership of 588, and 40% of committee members were female.

Each EMC was assigned the task of defining an environmental management action plan for the resources in their neighbourhood. In the course of defining these plans, a common set of issues emerged. First, in spite of efforts to clarify borders between neighbourhoods, there was continued confusion over which management committee was granted authority over particular resources due to the overlapping nature of resource use patterns for traditional definitions of neighbourhoods. Second, and related to this, there was poor integration between EMCs in different neighbourhoods; rules set by one group were not necessarily the same as those set by another group. This was especially problematic when the inconsistent rules were being applied to a resource for which there were overlapping claims. Third, there was apprehension about sanctioning members of one's own neighbourhood group, commonly expressed as a fear of curses. Fourth, there was no tangible incentive for members of the committees. In fact, they were being put in situations where they had to make decisions and risk angering their neighbours; if anything, this created an incentive to not be a member of the committee. Fifth, the legal status of the management committee was unclear, particularly as related to formal government rules and institutions.[1]

In response to these issues, between 1998 and 1999 MDP activities were mainly concentrated on bringing together representatives from different neighbourhoods to harmonize resource management protocols. The 29 management units with their distinct EMCs were clustered into 4 larger management units corresponding to the idea of a grazing area. Meetings were convened for EMCs and elderly leaders of each neighbourhood within a grazing area. Through a series of consultative meetings each of the larger grazing area units, EMCs and elderly leaders identified problems that had emerged in implementing the neighbourhood-specific management plans. Again, they noted that poor coordination and consultation among the EMCs had resulted in an escalation of internal community resource use conflict. In addition, they noted that neighbourhood-specific plans were still not able to cope with herders from other neighbourhoods who came into the area.

A harmonized natural resources management protocol detailing the procedures and penalties (based mainly on traditional customary laws) was drafted for enforcement at the grazing area level. Notable items in this protocol are how to:

1. Manage water resources;
2. Manage grazing land use by local residents;
3. Manage grazing land use by non-residents;
4. Manage use of tree species;
5. Establish rules over charcoal making;
6. Manage wild fires;
7. Manage movement of diseased livestock;
8. Develop communication mechanisms and dialogue with the local community;
9. Develop communication mechanisms and dialogue with neighbouring communities;
10. Develop communication mechanisms and dialogue with formal administrative structures;
11. Protect wildlife.

[1] The original intent was for community-level EMCs to function under the authority of the government-appointed District Environmental Officer (DEO). However, as the DEO position was vacant during the period of EMC formation, EMCs instead sought legal authority from other government departments and/or the chiefs of the respective village who had authority through the Chief's Act (Milimo *et al.*, 2002).

However, participants in these meetings argued that this harmonized management protocol did not address one of the main issues impeding environmental management efforts – insecurity. Insecurity had led populations to converge into more secure areas, leaving vast areas of the rangeland unused. Participants argued that addressing mismanagement without also addressing insecurity would ultimately be futile, as rest and rehabilitation of overused areas as called for in the management protocol required access to areas currently underutilized due to insecurity.

This placed the MDP staff in an interesting situation. By virtue of the wide geographical area they had worked in, they had built up trust and confidence with members of neighbouring ethnic groups who had little trust and confidence in one another. They had also developed a familiarity with the prevailing issues driving current conflicts. However, the programme was neither designed nor mandated to deal with conflict resolution. With some trepidation, the environmental management programme turned to issues of conflict management as they agreed with the assessment of the grazing area meetings – addressing conflict was a critical component of environmental management.

## 6.3 Inter-ethnic conflict management by peace committees

The Kenyan government's recognition that the capacity of formal security services to address inter-ethnic conflict was limited led to calls by government officials for cooperation between the administration, the police, the development agencies and the communities to reduce inter-ethnic conflict. Notably, the District Commissioner of Marsabit called on the local community in 1999 to cooperate with the administration to help address crime and insecurity in the district. Programmes targeted at addressing inter-ethnic conflict in Marsabit involving a variety of institutions – governmental, non-governmental and religious responded with vigour (Kenya Human Rights Commission, 2000).[2]

MDP began activities in this area between 1999 and 2002 with a series of consultative meetings to identify sources of conflict and define means to address conflict (Haro, 1999a,b). These meetings were held in collaboration with district-level government authorities, local government authorities, other non-governmental organizations (NGOs) active in the district, traditional authorities and EMC members. Resource management units were clustered around common resources regardless of their ethnic identity. The territory relevant to these clusters corresponded to areas used by the respective resource management units, regardless of ethnic and administrative boundaries. This latter characteristic was important as antagonistic groups used movement across administrative (including international) boundaries to attack other groups from an area they were not resident in. This would often lead to retaliatory attacks on the resident community, causing counterattacks, fuelling a cycle of violence which the formal administrative structure was not well designed to stop. For

[2] It should be noted that some of these institutions had been working on conflict resolution for many years prior to this statement by the government and the commencement of MDP activities aimed at conflict resolution.

the cross-ethnic peace initiative to succeed, it became necessary to bring on board all communities relevant to the use and management of a resource area, regardless of formal administrative boundaries.

Each group of community representatives was given an opportunity to describe its situation. It was an uninterrupted presentation to the facilitators in front of the other community representatives. The role of the facilitators was made clear to the participants. The facilitators were not judges or arbitrators to find out who was wrong and who right. Rather the role of the facilitators was confined to helping parties find amicable solutions since they were going to discuss their problems and they were the only ones who could determine acceptable solutions to the problems.

The community representatives were then divided into their ethnic groups and were requested to brainstorm on the conflict issues and their underlying causes. Each party was given time to present their deliberations. Over the course of these presentations, it began to be evident that participants realized they were all faced with similar problems. This in itself was an important insight that helped to tone down the tension between them. The following is a summary of the conflict issues that were identified:

1. Conflict over use and management of water;
2. Conflict over use of grazing areas;
3. Banditry, theft and murderous activities;
4. Unclear boundaries between neighbours;
5. Lack of cooperation among local leaders;
6. Lack of discipline among herders and warriors;
7. Overutilization of natural resources (trees and wildlife);
8. Lack of discipline among Kenya Police Reserves.

Having agreed on the pertinent issues and their causes, the participants were again divided into working groups defined by ethnicity to deliberate on and elaborate strategies to solve the identified problems. It was emphasized that the proposed solutions should be implemented within their own means, should be defined for themselves rather than for other groups to abide by, and should be acceptable to the other parties. Each group presented their propositions in plenary.

After long deliberations, the participants agreed on a set of rules that addressed the eight sources of conflict enumerated above. The agreement reached was then translated into all local languages and sent to all resource management units. It was then communicated to different categories of user groups (herders, warriors, women and children, chiefs and local elders, police reserves). For all EMCs falling into a common administrative location, members selected representatives for peace committees to oversee the implementation of the agreement.

## 6.4 Current status of environmental and conflict management efforts

Community management efforts for both environmental and conflict-related issues continue to this date. We briefly describe some of the accomplishments and challenges facing each type of effort.

Environmental management efforts have achieved some success. As noted above, management of trees was defined as an environmental issue to be addressed by both the intra-ethnic and inter-ethnic groups. Control over use of acacia species, doum palm and cedar has been particularly successful, as evidenced by the increased regeneration of these trees around the settlements in Marsabit district. Milimo *et al.* (2002) conducted an in-depth study of environmental issues in North Horr town where an EMC was active. They reported that by 1999 the majority of residents surveyed accepted the authority of the EMC and agreed that the committee is contributing to environmental rehabilitation in the area. They noted that the EMC had been active in encouraging natural vegetation regeneration for sand dune stabilization, operating tree nurseries and tree-planting activities.

In addition, efforts at reserving dry season grazing areas have been undertaken in a variety of neighbourhoods, although these have had limited success as of yet due to the drought of 2000/01 (these have been periods when use of the reserves was needed, so it is difficult to evaluate whether they will be set aside as reserves in non-drought years). A few other accomplishments have been that wildlife poaching has been reduced with respect to gazelles and one community has forbidden the use of plastic bags in local shops.[3]

Many of the issues identified at the intra-ethnic EMC meeting continue to be problems confronting effective community management of natural resources. Coordination among neighbouring EMCs continues to be a problem, as does the issue of non-neighbourhood residents being unaware of, or not respecting, resource management rules defined by the EMC. Concerns about the legitimacy of EMC rules within the neighbourhood also continue to be voiced. Although neighbourhoods designed these as rule-making bodies, the limited cultural precedent for such clearly defined rule-making groups leads some neighbourhood residents to refuse to accept the authority of the EMC as final. The formal authority of the EMC relies in most cases on the local chief, who may not always be willing to enforce EMC rules and sanctions.[4] In addition, EMCs are struggling to address the charge that their efforts are only addressing environmental issues around settlements and do not have an impact on the extensive grazing areas.

Inter-ethnic grazing cooperation has been successful at opening up areas unused previously due to insecurity, and has allowed formerly hostile groups to inhabit the same area. MDP staff suggests that prior to the programme's effort, roughly 60% of Marsabit rangelands were underutilized or unused due to insecurity, compared to the current estimate of 25%. Community-level research on land use changes conducted in 2002 in the Marsabit communities of North

[3] Beyond being visually unpleasant, residents of this area also complained that animals were eating the bags and dying because of ensuing intestinal complications.

[4] Milimo *et al.* (2002) provide another example of enforcement problems as they report that local police in North Horr were reluctant to punish local charcoal makers in spite of the fact that the local police and the EMC had signed a memorandum of understanding that obligated the police to arrest lawbreakers identified by the EMC as the police relied on charcoal as an energy supply.

Horr, Kargi and Logologo supports the contention that there are fewer rangeland areas unused due to insecurity.[5] Access to previously unused rangeland areas was described as particularly useful during the recent drought, as there were some areas of Marsabit district that received rain and had abundant pasture, which would have been unused in previous years due to insecurity. The presence of herders from different ethnic groups in the same area without any violence breaking out was seen as an important accomplishment that would not have occurred prior to the inter-ethnic meeting.

Follow-up meetings have been conducted after the initial 1999 workshops to monitor the progress of crisis prevention and conflict management activities. The purpose of these follow-up meetings is to assess the progress and also document how conflict issues between groups were managed by the established peace committees that could be improved in the future. Two main successes that were identified in these meetings are worth noting here. Revenge attacks did not occur when a Samburu herdsman murdered a Turkana herdsman in May 1999. After a long meeting, the Samburu elders agreed to compensate the family of the Turkana victim through the payment of livestock, and the case was resolved. Similarly, when a Gabra killed a Samburu boy in November 2001, the case was resolved without further violence. Again, after a meeting of elders, a payment in terms of livestock to the victim's family was agreed upon. While it would be preferable to prevent such loss of life from occurring in the first place, it was clear that the peace committees were able to prevent descent into a new spiral of violence.

A variety of issues confront conflict management efforts as they go forward. The most serious is that of collective retribution. In essence, if one of 'them' attacks 'us', 'we' attack 'them' in revenge regardless of whether the individuals attacked in revenge had anything to do with the initial attack. One manifestation of this is the complaint that EMCs are ill-equipped to handle raiders from outside the district who share a common ethnic identity with neighbourhood members when they launch an attack from within a neighbourhood with an EMC. Such attacks threaten the fragile peace between neighbourhoods who share resource areas but not a common ethnic identity. In a related fashion, conflict between ethnic groups in adjacent neighbourhoods is influenced by conflict outside of their area due to this principle of collective retribution. Cases such as the ongoing Turkana–Samburu conflict in Samburu district, and the Boran–Rendille conflict in the highlands of Marsabit threaten the peace between the groups that have managed to establish peace committees in other areas. News of an atrocity against one's own ethnic group by members of another ethnic group leads to calls for revenge attacks in any area where the two groups are in close proximity. Such spillover effects work the other way as well, as individuals and groups who have an interest in preventing alliances between other ethnic groups may actively seek to undermine the efforts at conflict resolution to advance their own interests.

[5] This research was conducted by one of the co-authors of this study (McPeak) as part of the USAID-funded GL-CRSP Pastoral Risk Management Project.

Conflict management committees also must confront ambiguities about the legitimacy of the resolutions reached by the peace committees in reference to formal administrative structures. While the terms of the agreement defined during the inter-ethnic meetings call for turning over an accused murderer to the police, local residents appear to have less confidence in the objectivity of formal legal structures in this area than they do in their own deliberations. In both cases of murder, traditional restitution in the form of livestock was agreed to by the traditional authorities in each community, and in the latter case the traditional agreement was reached on the condition that no formal legal proceedings would be pursued. The government of Kenya took an initial step towards addressing issues of legitimacy and legality of conflict management committees at a meeting convened by the Provincial Administration in Mado Gashe, Kenya, in 2001. The aim of this meeting was to harmonize the by-laws of different conflict management committees in different areas of northern Kenya and to put in place machinery to enforce these by-laws across administrative boundaries.

## 7 Conclusions

When there are multiple users who can exert a claim on a natural resource, management of the resource will almost inevitably require addressing conflicts arising from these multiple claims. Community management of natural resources does offer promise, but must explicitly consider the linkages between community management, environmental management and conflict management.

We have highlighted issues arising when there are multiple nested and overlapping definitions of a community that has claims on a given resource. Ambiguity in decision-making authority provides great flexibility for production in an uncertain environment, but also raises real challenges for resource management plans. The nested structure of social organization allows some possibility of achieving harmonization of management plans by calling together groups who share membership of a common larger social structure, as was illustrated by the MDP case. This allows harmonization of rules without formally allocating any one level of social organization exclusive decision-making authority, thus preserving aspects of flexibility without conflicting with cultural precedents. In addition, borders between communities may be ambiguous. Reconciling multiple claims within a single management protocol requires facilitating dialogue across decision-making authorities in different areas who have claims on a given resource. Dividing up rangeland into range management units based on vegetation type is often done in pastoral development plans, but the neat lines on the map may have little meaning to communities that have claims of varying strength on resources within different range units. Working with existing definitions of resource areas introduces some ambiguities, but also appears to offer some promise.

This study also illustrates that conflict management may be important for environmental management even if the conflict is not primarily due to contestation over a particular resource. While there is undoubtedly some element of resource competition involved in northern Kenya's insecurity, the cycle of violence and retribution has taken on a life of its own. Conflict management in this case was required to provide adequate security within which environmental management efforts could be undertaken.

Importantly, we find that communities are able to improve security by entering into dialogue with each other. What was required was facilitation to bring groups together and allow them to sort out their problems, and to define their own plan of action. It should also be noted that these efforts were successful due to the encouragement of formal administrative structures early in the process and the eventual ratification of the outcome by government institutions.

A different issue illustrated by this study is that community participation can lead development agents to become involved in issues that differ from their original programme focus. MDP's original focus on environmental management required programme staff to become involved in conflict management. Working together with the formal administration, other development agents, traditional leaders and community members, they were able to modify their programme to address conflict management directly. Again, we would stress that the support of the government agencies grew over time and contributed to the success of these efforts.

Overall, we find that improvement in the well-being of residents of pastoral areas is possible by working with pastoral communities and allowing them to define their own plans. The accomplishments to date of the MDP project have not resulted in a transformation of pastoral society, but rather built on the existing structure of pastoral society. Learning from the lessons of the MDP effort offers promise for efforts to improve environmental management and human welfare in other pastoral areas. This study has illustrated both the potential and the limits of community management of natural resources. We document that there are ways to establish and enforce rules in a common property production setting characterized by uncertainty without undermining flexibility in resource use patterns. We note that while it is well understood that resource scarcity can lead to conflict, it is also possible that conflict management can be an important element of addressing resource scarcity. Communities can identify solutions to both environmental degradation and insecurity if given facilitative support, which suggests that there is reason to be cautiously optimistic about pastoral development efforts which adopt this approach.

## Acknowledgements

This paper was prepared for the conference 'Reconciling Rural Poverty Reduction with Renewable Resource Conservation: Identifying Relationships and Remedies' Cornell University, Ithaca, New York, 1–3 May 2003. We thank participants at this conference for their helpful comments and criticisms. (Reprinted from *World Development* 33(2), 285–299. With permission from Elsevier.)

## References

Akabwai, D. (1992) Extension and livestock development: experience from among the Turkana pastoralists of Kenya. Pastoral Development Network Paper 33b. Overseas Development Institute, London.

Barnes, D. (1979) Cattle ranching in the semi-arid savannas of East and southern Africa. In: Walker, B. (ed.) *Management of Semi-arid Ecosystems*. Elsevier Scientific, Amsterdam, The Netherlands, pp. 9–54.

Barrett, C., Brandon, K., Gibson, C. and Gjertsen, H. (2001) Conserving tropical biodiversity amid weak institutions. *BioScience* 51(6), 497–502.

Brandstrom, P. (1985) Do we really learn from experience? In: Hjort, A. (ed.) *Land Management and Survival*. Scandinavian Institute of African Studies, Uppsala, Sweden, pp. 41–56.

Brown, L. (1971) The biology of pastoral man as a factor in conservation. *Biological Conservation* 3(2), 93–100.

Butcher, C. (1994) Extension and pastoral development: past, present and future. Pastoral Development Network Paper 37d. Overseas Development Institute, London.

Castro, P. and Nielsen, E. (2003) *Natural Resource Conflict Management Case Studies: An Analysis of Power, Participation and Protected Areas*. FAO, Rome.

Chambers, R. (1997) *Whose Reality Counts?* ITDG Publishing, London.

de Haan, C. (1994) An overview of the World Bank's involvement in pastoral development. Pastoral Development Network Paper 31b. Overseas Development Institute, London.

Doran, M., Low, A. and Kemp, R. (1979) Cattle as a store of wealth in Swaziland: implications for livestock development and overgrazing in East and southern Africa. *American Journal of Agricultural Economics* 61, 41–47.

Environment and Natural Resources Team (2002) *Nature, Wealth and Power: Emerging Best Practice for Revitalizing Rural Africa*. Sustainable Development Office (AFR/SD), USAID/USAID Development Experience Clearinghouse (DEC)/Development Experience System (DEXS).

FAO (2000) *Proceedings Electronic Conference on Addressing Natural Resource Conflicts through Community Forestry*. Prepared by D. Chandrasekharan. Community Forestry Unit, Forests, Trees and People Program, Forestry Department. FAO, Rome.

FAO (2001) *Integrating Conflict Management Considerations into National Policy Frameworks*. Community Forestry Unit, Forests, Trees and People Program, Forestry Department. FAO, Rome.

Fernandez-Gimenez, M. (2002) Spatial and social boundaries and the paradox of pastoral land tenure: a case study from postsocialist Mongolia. *Human Ecology* 30(1), 49–76.

Fratkin, E. (1991) *Surviving Drought and Development: Ariaal Pastoralists of Northern Kenya*. Conflict and Social Change series. Westview Press, Boulder, Colorado.

Galaty, J. (2002) Vue sur la violence: les frontières du conflit pastoral au Kenya. *Anthropologie et Sociétés* 26(1), 107–126.

Goodhue, R. and McCarthy, N. (2000) Fuzzy access: modeling grazing rights in sub-Saharan Africa. In: McCarthy, N., Swallow, B., Kirk, M. and Hazell, P. (eds) *Property Rights, Risk, and Livestock Development in Africa*. IFPRI, Washington, DC, pp. 155–190.

Haro, G. (1999a) Inter-ethnic Environmental Management Committees and Leaders Conference on Resource Use Dispute Management – Loiyangalani. MDP-GTZ in collaboration with the Kenyan Ministry of Agriculture. Marsabit, Kenya.

Haro, G. (1999b) Inter-ethnic Environmental Management Committees and Leaders Conference on Resource Use Dispute Management – Maikona. MDP-GTZ in collaboration with the Kenyan Ministry of Agriculture, ITDG and PISP. Marsabit, Kenya.

Homer-Dixon, T. (1991) On the threshold: environmental change as causes of acute conflict. *International Security* 16(2), 76–116.

Homer-Dixon, T. (1994) Environmental scarcities and violent conflict: evidence from cases. *International Security* 19(1), 5–40.

Ingles, A., Musch, A. and Qwist-Hoffman, H. (1999) *The Participatory Process for Supporting Collaborative Management of Natural Resources: An Overview*. FAO, Rome.

Jarvis, L. (1980) Cattle as a store of wealth: comment. *American Journal of Agricultural Economics* 62(3), 606–613.

Kellert, S., Mehta, J., Ebbin, S. and Lichtenfeld, L. (2000) Community natural resource management: promise, rhetoric, and reality. *Society and Natural Resources* 13, 705–715.

Kenya Human Rights Commission (2000) *The Forgotten People Revisited*. KHRC, Nairobi, Kenya.

Kratli, S. and Swift, J. (1999) *Understanding and Managing Pastoral Conflict in Kenya: A literature review*. Institute for Development Studies, University of Sussex, Sussex, England.

Lind, J. (2002) Report of the consultative session and regional conference on the ecological sources of conflict in sub-Saharan Africa. *Eco-conflicts* 2:1. African Center for Technology Studies, Nairobi, Kenya.

Lind, J. and Sheikh, N. (2001) Armaments, Environments: Small Arms and the Control of Natural Resources. *Eco-conflicts* 1:3. African Center for Technology Studies, Nairobi, Kenya.

Lind, J. and Sturman, K. (eds) (2002) *Scarcity and Surfeit: The Ecology of Africa's Conflicts*. Institute for Security Studies, Pretoria, South Africa.

Lipscomb, J. (1955) *White Africans*. Faber & Faber, London.

Lusigi, W. (1981) *Combating Desertification and Rehabilitating Degraded Production Systems in Northern Kenya: An IPAL Case Study and Summary of Results So Far*. UNESCO, Nairobi, Kenya.

Lusigi, W. (1984) *Integrated Resource Assessment and Management Plan for Western Marsabit District, Northern Kenya*. UNESCO, Nairobi, Kenya.

McPeak, J. (2003) Analyzing and addressing localized degradation in the commons. *Land Economics* 79(4), 515–536.

Milimo, B., Olukoye, G. and Moindi, D. (2002) *The Human Dimension of Desertification with Special Reference to People's Participation in Dune Stabilization Measures in North Horr (Northern Kenya)*. GTZ, Eschborn/Germany.

Moore, K., Bertelsen, M., Diarra, L., Kodio, A., Cisse, S. and Wyeth, P. (2000) Natural resource management institution building in the decentralizing context of West Africa: the SANREM CRSP approach. Working Paper 01–02. Office for International Research and Development, Virginia Tech, Blacksburg, Virginia.

Moris, J. (1998) Under three flags: the policy environments for pastoralists in Ethiopia and Kenya. SR/ GL-CRSP Pastoral Risk Management Project Technical Report 04/99. Utah State University, Logan, Utah.

Niamir-Fuller, M. and Turner, M. (1999) A review of recent literature on pastoralism and transhumance in Africa. In: Namir-Fuller, M. and Turner, M. (eds) *Managing Mobility in African Rangelands: The Legitimization of Transhumance*. Intermediate Technology Publications, London, pp. 18–46.

Nugent, J. and Sanchez, N. (1999) The local variability of rainfall and tribal institutions: the case of Sudan. *Journal of Economic Behavior and Organization* 39, 263–291.

Oba, G. (1985) Local participation in guiding extension programs: a practical proposal. *Nomadic Peoples* 18, 27–45.

Oba, G. (1992) Traditional grazing systems of the Rendille and Ariaal pastoralists: changing strategies and options in the southern district of Marsabit, northern Kenya. Marsabit Development Programme, GTZ, Marsabit, Kenya.

Odhiambo, M. (2000) *Oxfam Karamoja Conflict Study: A Report*. Oxfam, Kampala, Uganda.

O'Leary, M. (1987) Changing responses to drought in northern Kenya: the Rendille and Gabra livestock producers. In: Baxter, P. (ed.) *Property, Poverty and People: Changing Rights in Property and the Problems of Pastoral Development*. Department of Social Anthropology and International Development Center, University of Manchester, Manchester, England.

Ostrom, E. (1990) *Governing the Commons: The Evolution of Institutions for*

*Collective Action*. Cambridge University Press, Cambridge.

Ostrom, E. (1992) *Crafting Institutions for Self-Governing Irrigation Systems*. Institute of Contemporary Studies, San Francisco, California.

Pratt, D. and Gwynne, M. (eds) (1977) *Rangeland Management and Ecology in East Africa*. Hodder & Stoughton, London.

Pratt, D., Le Gaal, F. and de Haan, C. (1997) Investing in pastoralism: sustainable resource use in arid Africa and the Middle East. World Bank Technical Paper 365. World Bank, Washington, DC.

Ribot, J. (2002) *Democratic Decentralization of Natural Resources: Institutionalizing Popular Participation*. World Resources Institute, Washington, DC.

Robinson, P. (1985) *Gabra Nomadic Pastoralism in Nineteenth- and Twentieth-Century Northern Kenya: Strategies for Survival in a Marginal Environment*. Unpublished PhD dissertation. Northwestern University, Department of History, Evanston, Illinois.

Sandford, S. (1983) *Management of Pastoral Development in the Third World*. Wiley, New York.

Schlee, G. (1989) *Identities on the Move: Clanship and Pastoralism in Northern Kenya*. Gideon Were Press, Nairobi, Kenya.

Schlee, G. (1990) Policies and boundaries: perceptions of space and control of markets in a mobile livestock economy. Sociology of Development Research Center, Africa Program Working Paper 133. Universität Bielefeld, Bielefeld, Germany.

Schwartz, J., Shaabani, S. and Walther, D. (1991) *Range Management Handbook of Kenya*, Vol. II-1: *Marsabit District*. Ministry of Livestock Development, Nairobi, Kenya.

Scoones, I. (1995) New Directions in Pastoral Development in Scoones, I. (ed.) *Living with Uncertainty: New Directions in Pastoral Development in Africa*. Intermediate Technology Publications, London, pp. 1–36.

Sobania, N. (1979) *Background History of the Mt. Kulal Region of Kenya*. UNESCO, Nairobi, Kenya.

Spencer, P. (1973) *Nomads in Alliance: Symbiosis and Growth among the Rendille and Samburu of Kenya*. Oxford University Press, Oxford.

Sylla, D. (1995) Pastoral organizations for uncertain environments. In: Scoones, I. (ed.) *Living with Uncertainty: New Directions for Pastoral Development in Africa*. Intermediate Technology Publications, London, pp. 134–152.

Turner, M. (1999a) Conflict, environmental change, and social institutions in dryland Africa: limitations of the community resource management approach. *Society and Natural Resources* 12, 643–657.

Turner M. (1999b) The role of social networks, indefinite boundaries and political bargaining in maintaining the ecological and economic resilience of the transhumance systems of Sudano-Sahelian West Africa. In: Namir-Fuller, M. and Turner, M. (eds) *Managing Mobility in African Rangelands: The Legitimization of Transhumance*. Intermediate Technology Publications, London, pp. 97–123.

van den Brink, R., Bromley, D. and Chavas, J.-P. (1995) The Economics of Cain and Abel: agro-pastoral property rights in the Sahel. *The Journal of Development Studies* 31(3), 373–399.

Vedeld, T. (1998) State law versus village law: law as exclusion principle under customary tenure regimes. In: Berge, E. and Stenseth, N. (eds) *Law and the Management of Renewable Resources*. Institute of Contemporary Studies, San Francisco, California.

Walker, B. (1979) Introduction. In: Walker, B. (ed.) *Management of Semi-arid Ecosystems*. Elsevier Scientific, Amsterdam, The Netherlands, pp. 3–5.

World Bank (1985) The smallholder dimension of livestock development. Operation and Evaluation Department, Internal Report. The World Bank, Washington, DC.

# 8 The Unfulfilled Promise of Microfinance in Kenya: The KDA Experience

S.M. Osterloh and C.B. Barrett

## 1 Introduction

Microfinance offers promise for alleviating poverty by providing financial services to people traditionally excluded from financial markets. Small-scale loans can relieve capital constraints that might otherwise preclude cash-strapped entrepreneurs from investing in profitable businesses, while savings services can create opportunities to accumulate wealth in safe repositories and to manage risk through asset diversification. While this promise of microfinance is widely touted, it is infrequently subject to careful evaluation using detailed data.

This chapter examines the extension of microfinance services to people in Kenya. Using data collected from 17 financial service associations (FSAs) founded by the Kenya Rural Enterprise Programme (K-REP) Development Agency (KDA), we explore the intricacies of microfinance institutions (MFIs) emerging in these challenging environments.

Similar to cooperatives, FSAs mobilize local resources, capitalize upon local information and wealth, and tie it back into the local economic system by investing in borrowers who are members of the FSA community. Human resources too are local, as staff and board members are recruited from the village and trained by KDA in FSA administration, accounting and portfolio management. FSAs introduce scale-appropriate investment instruments in equity as well as in savings. By presenting accessible opportunities to hold assets in a different and hypothetically safer form, FSAs can, in principle, provide a safe repository for savings as well as start-up capital to potential entrepreneurs, who otherwise might be prevented from engaging in business due to binding liquidity constraints. If they can indeed foster asset diversification and the creation of business opportunities, FSAs have a potentially pivotal role to play in regions struggling to achieve economic growth. But given the experience we document below, achieving this promise seems to require some changes in design and practice.

FSAs face formidable challenges in Kenya, including poor infrastructure, formal banking systems inaccessible to the poor, lack of savings alternatives, and pockets of politically and economically marginalized populations. Moreover, the promise of decentralized decision making to obviate information problems, to mobilize local wealth and to encourage community-based investment can fall prey to power relations that pervade poor communities. The promise of microfinance, including the FSA model, is undeniable; the reality of its experience nevertheless demands careful scrutiny.

This chapter investigates whether or not the provision of KDA's microfinance services in Kenya realizes this potential to help participants mitigate risk, foster profitable investment and improve welfare. Ultimately, we find that despite their stated pro-poor mission, KDA FSAs typically bypass the poorest members of the communities they serve and implicitly foster regressive wealth distribution from lower-income, non-borrowing members to higher-income members more likely to take out, and default on, large loans. An important lesson to learn is that without careful analysis of the patterns of benefits reaped and the bearers of costs borne, we cannot accept at face value that an initiative is 'pro-poor'. The KDA experience appears to be one of unfulfilled promise.

## 2 Microfinance in Brief

Microfinance refers to small-scale financial services such as cash loans, money transfers, direct deposits, savings and insurance made accessible primarily to the poor. Two prominent features of successful MFI building are group lending and savings (Yaron, 1994).

Popularized by the Grameen Bank in Bangladesh, group lending refers to the practice of issuing loans to individual members of small, homogeneous groups. Participants self-select into groups, which collectively guarantee loans issued to their members. All members are barred from further access to credit in the case of default by one group member, providing strong incentives for the group to ensure repayment by each individual borrower.

Stiglitz' (1990) seminal work on peer selection and monitoring argues that joint liability reduces informational asymmetries between borrower and lender. Without indicators signalling creditworthiness, such as established credit histories and regular wage salaries, outside lenders protect against the threat of default by charging a high rate of interest or requiring significant collateral, irrespective of credit risk heterogeneity among borrowers. The lender's inability to observe the borrower's risk type yields the standard example of adverse selection.

In contrast with external formal sector lenders, a community shares a great deal of knowledge about its members. Joint liability harnesses this familiarity to circumvent the problem of asymmetric information. Participants affiliate with others of similar risk profiles because the group is liable for the loan repayments of all members. The threat of sharing liability on a defaulting member's loan moderates the problem of adverse selection as groups rationally exclude potential members with high levels of *ex ante* risk. Using community knowledge, interdependent groups reduce the lender's *ex ante* risk exposure in two ways: by excluding high-risk types from joining into their group and by approving only those loans whose risk they are willing to share within the group. Contractual interdependence through group lending not only reduces *ex ante* risk, but also moderates *ex post* risk. Intensive peer monitoring of the borrower after loan disbursement diminishes moral hazard, or diversion of the loan towards investments riskier than its agreed-upon purpose. Borrowers repay in a timely and efficient manner to safeguard the group's continued access to future loans. By reducing the risks of asymmetric information and moral hazard, groups enable the lender to afford lower interest rates and reduce, or eradicate, collateral requirements. As the poor lack vehicles for wealth creation, the ability of group lenders to circumvent stringent collateral requirements via joint liability is a key feature rendering microfinance accessible to the poor.

Group lending reduces informational asymmetries and increases borrowers' willingness to repay through informal sanctions. Besley and Coate (1995) show that group-levied social sanctions, including peer pressure, loss of social prestige and social isolation, can improve rates of loan repayment. Joint liability transfers certain fixed costs of small-scale lending, including screening, contract, monitoring and enforcement, from the bank to the group, rendering financially viable the administration of small loans in rural areas to people traditionally regarded as high risk. Although joint liability imposes some non-trivial costs upon participants, including frequent group meetings and limited access to individual credit, members of groups continue to seek loans. Empirical evidence generally supports the superiority of group lending repayment rates over repayment rates of loans issued to individuals (Adams and von Pischke, 1992; Yaron, 1994; Conning, 1999; Woolcock, 1999).

Savings too emerges as a critical component of MFI success by serving as collateral on loans and 'introduc(ing) and enhanc(ing) financial discipline among inexperienced, first-time, small-scale borrowers' (Yaron, 1994, p. 52). Microcredit is explicitly costly, with effective annual interest rates of up to 130%, while savings is relatively cheap, leading Yaron and others to speculate that savings facilities can serve a greater number of clients than lending services. They further claim success in mobilizing savings as the optimal, and in some cases unique, path towards MFI financial self-sustainability. Well-established empirical evidence supports this perspective (Hulme and Mosley, 1996; Christen, 1998; Hollis and Sweetman, 1998; Morduch, 2000).

## 3 Overview of KDA

Founded in 1984 by World Education, Inc., K-REP was designed to provide credit and technical assistance to financial services NGOs (Pederson and Kiiru, 1997). Pederson and Kiiru report that in 1989, K-REP's Board of Directors expanded services to include a minimalist lending programme through existing rotating savings and credit associations (ROSCAs) and newly formed joint liability groups. From 1991 to 1995, the two programmes enjoyed a collective repayment rate of 94%. Inspired by this high rate of repayment and the 1996 concept paper by Dr A. Jazayeri, K-REP introduced the Financial Services Association (FSA). After 4 years of rapid expansion, K-REP underwent restructuring in 2001, resulting in three sister organizations, K-REP Bank, K-REP Consulting and K-REP Development Agency, the last of which is devoted exclusively to the management of the FSAs. As of December 2000, KDA's 52 FSAs had attracted a total of 20,356 members who had received over 6680 loans, valued at KSh 33,259,000.[1] By 2001, KDA had reached the upper bounds of its organizational capacity, and began eschewing donors' offers for new FSA funding in order to refocus attention from FSA expansion to FSA strengthening.

To address the severely limited financial services available in areas traditionally ignored by commercial banks, Jazayeri (1996) modelled the FSA as a self-reliant, small-scale financial institution catering to a community's niche needs by harnessing local equity capital, thereby releasing the institution from dependence upon the goodwill of external funding agencies. A survey of ten prominent MFIs in Africa and Asia by Zeller *et al.* (1997) concludes that limiting the role of donors to funding start-up costs increases the chances of MFI financial viability. KDA limits its FSA commitment to providing start-up capital for non-productive assets; the initial provision of operational materials; training in bookkeeping, marketing, and management for the FSA officials, staff and members; and a yearly audit service to check the books, calculate share value and provide technical assistance. KDA grants these goods and services to the FSA, and in addition hires and trains a field coordinator (FC), a professional with an accounting background, native to the region. FCs are assigned several geographically proximate FSAs and assist in raising share capital, while helping the fledgling FSA with addressing technical issues.

In the FSA model, the institution is owned and managed by the community, and targets the poorest 20% of the population (Pederson and Kiiru, 1997). Locals buy FSA shares at KSh 300 per share. Share ownership confers FSA membership, which includes access to loans and savings services. The value of the shares sold generates a loan fund, redistributable amongst members via credit, and KDA does not augment share capital. Each member has one vote in the annual election of the FSA Board of Directors, comprising eight members with staggered 3-year terms. The Board of Directors hires staff members and applies local knowledge about FSA members and their creditworthiness to screen loan applications. Meeting once a month, the Board evaluates new loan requests and reviews current loan repayment status. Responsibility for the collection of bad debt lies primarily with the Board of Directors. Monetary remuneration of the Board is infrequent, but not prohibited by the by-laws of the FSA. Incentive to screen out bad loan risks and to recover delinquent loans arises from the Board's vested interest as investors in the FSA and the social (and legal) responsibility conferred by election, as opposed to salaries or wages.

FSA by-laws allow members to apply for loans up to four times the value of their shareholdings, thus rewarding investment in the bank, while at the same

[1] The average exchange rate in 2000 was approximately KSh 76/US dollar.

time limiting idiosyncratic risk exposure. Loan applications require personal information as well as loan use details, and make explicit the consequences of loan default, which include the imposition of stiff penalties, the seizure and subsequent sale of collateral and collection pressure applied to loan guarantors. In addition, members are encouraged to deposit savings in the FSA, up to ten times the value of their shareholdings. In contrast to the sale of shares, monies deposited as savings cannot be invested; thus, the FSA faces a 100% reserve requirement on deposits. Because of this restriction on the employment of deposits, savings earn no interest payments.

Protection of equity investment encourages FSA members to monitor borrowers and ensure prompt repayment. Share values increase as FSAs earn profits from interest payments and penalties collected, and fall when loans default. All shareholders share an equity incentive to stay informed about the status of FSA loans. Facilitating the exchange of information between FSA staff, the Board of Directors and shareholders, annual general meetings (AGMs) make explicit the current share value and publicize borrowers in poor standing, generating the threat of social sanctions imposed on defaulters by the shareholding community. The ideal of stringent application screening combined with intensive peer monitoring encouraged Jazayeri to predict the average FSA to attain total administrative and financial sustainability in 2 years.

## 4 Data and Methods

The objective of this section is to explain the methods used in data collection and analysis to address the following questions:

**1.** Is KDA reaching its target population? Who purchases shares and joins the FSA?
**2.** To what extent do members utilize FSA services? Which members reap rewards from FSA services?
**3.** Do FSAs achieve financial sustainability and the development objectives of KDA and its donors?

To answer these questions, we visited 14 of KDA's 52 FSAs over the course of 8 months from January to August 2000. Of these, five were located in the Kwale district of Coast Province (Lunga Lunga, Tiwi, Msambweni, Mwaluphamba and Kikoneni), four in the Migori district of Nyanza Province (Rongo, Mbita, Magunga and Karungu), and five in the Marsabit district of Eastern Province (North Horr, Gabra Scheme, Kalacha, Korr and Badha Hurri). The provinces were selected by stratified purposeful sampling, maximizing geographic diversity and selecting economic systems and population density as the subgroups of interest. Located in the far south-east of the country, Kwale is primarily an agricultural zone with a mid-range population density of 57/km$^2$ (CBS, 2001). Migori, sitting in the far south-west corner adjacent to Lake Victoria, boasts the highest population density in the country at 120/km$^2$ with an economy based upon a mixture of fishing and cash crop production (CBS, 2001). Marsabit in the northern section of the country features arid lands, a pastoralist economy and the lowest population density in Kenya at 2/km$^2$ (CBS, 2001).

Site-specific enumerators collected data from FSA staff, officials, members and FSA non-members. Information on demographics, wealth, income, savings and loans access and use, and FSA participation was solicited. Each enumerator attempted to interview the entire Board of Directors, 25 FSA members and 25 non-members. The sampling frame was established by stratifying each site's population into clusters of members and non-members. The member interviewees were selected at fixed intervals with a random starting point from the complete roster of adult shareholders. Upon completion of a member survey, enumerators scoured the immediate vicinity of the interview for a non-member of the same age and gender as the member respondent to participate. In this quasi-experiment, non-members were not randomly sampled but rather informally matched with randomly sampled FSA members.

In the Eastern Province, the members herding animals far from town were excluded from the sampling frame. Only town dwellers and members located not further than 10 km from town were included in the sampling frame. Truly nomadic pastoralist members were excluded, due to lack of transportation and roads, and lack of accurate information about the whereabouts of pastoralists in search of forage and water. This selection may cause a bias in the data as the sample of northern members surveyed is not entirely random. However, few members were excluded due to inaccessibility, with the exception of Kalacha FSA, where one-third of the originally sampled members were summarily excluded due to inaccessibility. In total, 17 staff, 108 Board, 282 member and 292 non-member surveys were collected from 14 sites.

## 5 FSA Performance

We begin with a cross-sectional analysis of data derived from monthly monitoring reports submitted to KDA headquarters by FSAs. The cross-sectional reports date to November and December 2000, and include one observation from 51 of KDA's 58 FSAs (as of August 2004), located in 16 districts across Kenya. It should be noted that the data represent claims made by the FSAs, and were not verified. While data collected from the field often contradict the monthly monitoring report information, these are the only data available to work with and are therefore used as second-best indicators.

The data as reported to KDA and updated by field observation were used to estimate the effect of FSA characteristics upon share value using the following regression model.

Sharevalue = *f*(Headcount, North, LGen, Members, Members$^2$, Loans, Loans$^2$, Savers, Savers$^2$), where Sharevalue is a continuous variable representing the share value in KSh as calculated between the period of November 2000 and March 2001. Although in principle this variable is bounded from below by zero, none of the FSAs exhibited complete erosion of share value, so we use a simple ordinary least squares estimator. Recall that members across all FSAs purchase shares at KSh 300 and that money is pooled together to form share

**Table 8.1.** OLS estimates of determinants of FSA share value.

| Variable | Coefficient estimate | Robust standard error |
|---|---|---|
| Headcount | 1.481** | 0.668 |
| North | −99.834** | 41.908 |
| LGen | −12.745* | 6.481 |
| Members | 0.174** | 0.057 |
| Members$^2$ | −0.001*** | 0.000 |
| Loans | 0.920** | 0.401 |
| Loans$^2$ | −0.001 | 0.001 |
| Savers | −0.193 | 0.181 |
| Savers$^2$ | −0.001 | 0.000 |
| Constant | 203.425*** | 52.732 |

$n = 51$; $r^2 = 0.718$.
Notes: ***signifies estimates that are statistically different from zero at the 1% level; **at the 5% level; and *at the 10% level.
Since FSAs in the same regions tend to share characteristics, robust standard errors are used to take into account heteroskedasticity and inter-region correlation (White, 1980).

capital, which is redistributed among borrowing members in the form of loans. Share value reflects the health of the loan portfolio. Repaying loans replenishes share capital, collection of loan interest and fees augments share capital, and losses due to loan default reduce share capital. Share capital on hand at the FSA plus the expected return from loans not in default is divided among the number of shares issued to calculate share value.

How is FSA share value affected by poverty in the region? (See Table 8.1.) Headcount captures the share of the population whose consumption falls below the poverty line, defined as the consumption of KSh 1239 per adult equivalent in rural locations per month, and KSh 2648 in urban locations (Government of Kenya, 2000). This variable is drawn from the compilation of poverty measures by the Central Bureau of Statistics (CBS, 2003) of the Kenyan Ministry of Planning and National Development. District averages are used for FSAs in locations not reporting poverty statistics. Of the 38 FSAs for which poverty statistics are available, 61% are located in poorer locations relative to the district average. The expected impact of poverty in the FSA area is unclear. On the one hand, increased poverty might decrease the profitability of FSAs as business opportunities in poorer areas tend to be less lucrative. On the other hand, greater poverty may mean fewer alternative sources of financial services, reducing competition and making the FSA more profitable, thereby enhancing share value.

The binary variable North takes on value 1 if the FSA is one of seven located in the eastern or north-eastern districts, and zero otherwise. The variable North represents FSAs serving pastoralists in transition. Lack of business opportunities, coupled with high risks and poor infrastructure in this region, leads to a negative expected coefficient estimate associated with share value.

Loan Generation (LGen) is a count variable quantifying the distinct rounds of lending which occurred between the opening of the FSA and December

2000. For FSAs which had yet to offer loans, this variable equals zero. Due to capital constraints, the next round of loans cannot be disbursed until the majority of the previous wave has been repaid. Although loans are approved in the interim between disbursement and repayment, new loans cannot be issued until share capital returns to the FSA. Rather than a semi-continuous process, loans tend to be issued at 6-month intervals. In the case where loans are repaid in a timely manner with interest, additional LGen builds equity and should have a positive effect upon share value. If, however, credit repayments are low, share values erode over time and the coefficient estimate on the LGen variable will be negative. Overall, 67% (34 of 51) of FSAs reported share values higher than the purchase price of KSh 300, and 25% (13 of 51) reported share value loss. Note that without disbursing loans (LGen equals zero), share value can be no higher or lower than KSh 300.

Members (the number of FSA shareholders) and $Members^2$ (the number of members squared) capture the prospective non-linear relationship between membership and FSA profits. Although expansion of the FSA by member recruitment increases the loan fund, it may simultaneously tax the FSA community's ability to self-regulate through screening and peer monitoring. If the benefits of additional loan capital outweigh the added costs of monitoring new members, membership should increase expected share value. If, on the other hand, the costs of expansion outweigh the benefits of additional loan capital injected by the sale of shares to new members, the estimated effect on Sharevalue could be negative. Field observations suggest that FSA boards and staff do not always rigorously evaluate loan applications and that members are not particularly aware of their peers' loan status. Therefore, we hypothesize that expansion of the FSA community might be beneficial, up to a point, in expanding the pool of loanable funds, but that expanded membership gradually diminishes the FSA's capacity to carefully manage the loan fund. This would imply positive and negative estimated coefficients for the Members and $Members^2$ variables, respectively.

Loans and $Loans^2$ represent the number of loans disbursed and that quantity squared, respectively. The logic of this quadratic specification is similar to that for membership. A high count of loans disbursed, controlling for LGen, likely signals great demand for credit due to profitable local business opportunities. This should be associated with increased share values, as reflected in positive estimated regression coefficients on at least the first of these variables.

As mentioned previously, a strong emphasis upon savings typifies many successful microfinance programmes. In KDA's model, however, FSAs are prohibited from productively employing savings except as deposits in commercial banks, a linkage which none of the sampled FSAs had yet established. From the FSA's perspective, savings represents a liability as it must be administered and safeguarded at the opportunity cost of additional share capital, which could theoretically earn a positive rate of return through efficient loan repayment. Although regulations render savings deposits a deadweight liability, savings can none the less serve as loan insurance or collateral, to be confiscated in the event of loan default, or less drastically, to draw upon for loan repayment. In addition, some FSAs have implemented savings withdrawal fees, to pay for the fixed costs of transactions. Savers tallies the number of members who have ever saved at the FSA, and $Savers^2$ is the number of savers squared. If the role of savings in the FSA is indeed a key determinant to success as predicted in the literature and if profits increase in savers at an increasing rate, the regression should yield positive coefficient estimates for these savings variables.

Contrary to our expectation, the incidence of poverty is positively associated with FSA share value. This result suggests that poorer people are more likely to repay FSA loans, an idea we develop later. With fewer financial alternatives available, people from more impoverished regions have greater incentive to repay loans, so as to remain in good standing with the FSA. Pearce and Helms (2001) note that FSA interest rates are uncompetitive compared with other savings and credit cooperatives and ROSCAs, suggesting that FSAs might struggle in more competitive financial services markets. Among the 14 FSAs sampled, the bivariate correlation coefficient between the poverty headcount index and the number of non-FSA financial alternatives accessed within the last year by respondents averaged across FSA site was –0.27. If this correlation holds across districts and the number of financial services used by respondents does decrease as the poverty incidence increases, the positive coefficient estimate associated with the Headcount variable might substantiate Pierce and Helms' claim.

The coefficient estimate associated with the North dummy variable is negative and significant, suggesting that northern FSAs are roughly half as profitable ((203.425–99.834)/203.425) as their down-country counterparts, all else equal. Poor spatial market integration, lack of infrastructure, high levels of covariate risk (e.g. due to drought), a relatively underdeveloped cash economy and a dearth of non-livestock investment opportunities all likely contribute to this poorer average performance. Controlling for the headcount measure of poverty, the North variable reflects a fixed effect for the relatively isolated region with modest diversification of commercially viable opportunities.

This is reflected in transport costs and their implications for the cap on loan size, four times shareholdings. The dismal state of roads in Marsabit district (Eastern Province), coupled with the lack of reliable transportation, results in long, arduous, expensive journeys into the district capital from settlements elsewhere in the district. In most other locations, the smallest loan size accessible to members holding one share (KSh 1200) is sufficient to purchase sundries (e.g. soap, spices, tea) for petty trade and to cover the fixed transactions costs of travelling to town to buy these supplies. In four of the five northern FSAs, however, loans of KSh 1200 are insufficient to cover even the round-trip transport costs to Marsabit town, the nearest source for traders' provisions, leaving nothing for the acquisition of inventory. The negative estimated coefficient on the North variable may thus reflect regional differences in minimum efficient scale of lending given regional variation in the costs of commerce.

The greater the number of times the loan fund is turned over, the lower the return on investment, as indicated by the LGen's significant, negative coefficient. Given that the costs of operation do not vary greatly over the lifetime of the FSA, this loss reflects principal default rates not adequately covered by income from interest and penalty payments, sale of passbooks to new members, and services fees on loan applications and savings withdrawals. Since FSA income not related

**Table 8.2.** Average membership and loans issued by district.

| District | Average membership across FSAs in district | | | Average number of loans issued |
|---|---|---|---|---|
| | Mean | Minimum | Maximum | |
| Bomet/Buret | 1170 | 428 | 1979 | 255 |
| Busia/Teso | 271 | 152 | 346 | 42 |
| Garissa | 420 | 404 | 436 | 138 |
| Kilifi | 274 | 170 | 415 | 87 |
| Kwale | 317 | 287 | 400 | 74 |
| Machakos/Kitui | 253 | 158 | 402 | 223 |
| Makueni | 232 | 195 | 285 | 12 |
| Marsabit | 188 | 109 | 390 | 35 |
| Meru South | 567 | 567 | 567 | 85 |
| Suba/Migori | 353 | 279 | 433 | 95 |
| Taita/Taveta | 336 | 56 | 616 | 301 |

to loan servicing represents only 4–17% of FSA cash flow among sampled FSAs in operation for 1 year or more, repayment rates must necessarily be high in order to ensure preservation of share value in FSAs under present interest rates and fee schedules.

The negative estimated coefficient on LGen does not bode well for the future of FSAs. Controlling for other location and FSA characteristics, if share value erodes at a constant rate of KSh 12.75 per LGen, a share purchased during FSA inception loses all value in approximately 15 rounds of loans, or roughly 12.5 years. This is plainly unsustainable.

The coefficient estimates for Members and Members$^2$ are significant, positive in the linear term and negative in the quadratic term, suggesting that returns to membership increase, but at a diminishing rate up to some optimal membership level, which we estimate as 760 members. The estimated optimal membership size is significantly larger than the average memberships of all FSA districts, with the exception of Bomet/Buret, home to four FSAs with memberships larger that 760 (Table 8.2). It thus appears that most FSAs have not yet grown to optimal size. Of course, if share value is eroding, as reflected in the negative coefficient estimate on LGen, it becomes difficult to attract new members, as FSA membership appears a money-losing proposition. Reaching optimal scale, therefore, likely depends on improvement in lending design and performance to stem share value degradation.

The coefficients associated with Loans and Loans$^2$ yield a similar picture, although the negative point estimate associated with the quadratic term is not statistically significantly different from zero at conventional levels. Share value increases at a decreasing rate in the number of loans issued, holding membership and LGen constant with an optimum at 569 loans (relative to optimum membership size of 760 persons), significantly larger than the average number of loans issued across FSAs in any district (Table 8.2). Moreover, since the second-order term's coefficient estimate is not significantly different

**Table 8.3.** Savers and savings by district.

| Districts | Savers as % of members | Savings as a % of share capital |
|---|---|---|
| Taita/Taveta | 80 | 54 |
| Marsabit | 9 | 5 |
| Machakos/Kitui | 69 | 163 |
| Suba/Migori | 96 | 68 |
| Bomet/Buret | 29 | 53 |
| Kilifi | 52 | 32 |
| Kwale | 53 | 47 |
| Meru South | 54 | 6 |
| Busia/Bungoma/Teso | 56 | 180 |
| Garissa | 26 | 33 |
| Makueni | 57 | 148 |
| Average across all districts | 53 | 72 |

from zero, this result suggests that the most profitable FSAs maximize lending volume each generation, signalling that the effectiveness of screening and monitoring does not seem to decrease with scale of lending once one controls for generation and membership size.

Contrary to the claims one finds throughout the literature, savings does not lead to increased FSA profitability. This result could be due to the stringent reserve requirement demanded of FSAs, not observed in other microfinance models where institutions are allowed to lend out a percentage of savings. Although some FSAs may benefit from savings through the collection of withdrawal fees, most deem savings a burdensome transfer of liability from members to the institution. It would seem that KDA would be well advised to re-examine the role of savings in the FSA.

Both the percentage of members who save and savings expressed as a percentage of share capital vary greatly across districts (Table 8.3). Marsabit and Meru South demonstrate particularly low rates of net savings to share capital, while Machakos/Kitui, Makueni and Busia/Teso have particularly high rates of net savings. When we consider the number of savers, we find that the two districts in the north, Marsabit and Garissa, exhibit the lowest proportion of savers among their shareholders, suggesting that if there is a latent demand for savings among pastoralists, it is not observed by northern FSAs.

## 6 Individual FSA Membership and Product Use Decisions

Our results to this point indicate that FSAs are not fully living up to their promise. Share values erode as the institutions mature, reflecting earnings insufficient to cover loan defaults. We have established, however, that FSA performance varies markedly across locations. Profitability varies with FSA characteristics such as membership size and loan volume, as well as geographic factors, including the extent of competition from other financial service providers, as proxied by the local headcount poverty rate, the quality of infrastructure and the local importance of covariate risk (captured by the North dummy variable).

The characteristics investigated describe the share values of all members of FSAs. Since microfinance is a vehicle for assisting the poor, it is desirable to consider at a more disaggregated level who joins FSAs, especially in light of steadily eroding share values, which suggest that the average member loses wealth by joining. Further, conditional on having selected an FSA, who purchases multiple shares (securing access to larger loans up to four times their shareholdings), who borrows from the FSA and who repays their loans? These patterns of individual behaviour determine the distributional effects of the FSA within Kenyan communities. The individual household survey data we collected offer an uncommon opportunity to address these key questions.

## 7 FSA Membership

Do FSAs serve the poorest residents in Kenya? Relatively few people with access to an FSA actually become members. Using 1999 Census data, membership in December 2000 ranges from less than 1% of the adult population in the Suba/Migori districts to just over 14% in Bomet/Buret districts. Membership rates are highest in towns such as Korr and North Horr (11% and 17%, respectively), as opposed to rural areas with dispersed populations such as Mwaluphamba (less than 2%) and larger urban centres like Mbita (less than 1%), where several alternative providers of financial services exist.

Given the relatively modest membership rates within jurisdictions, who joins the FSA? This question is explored using the following probit regression model.

Member = *f*(YQuint1–4, WQuint1–4, Primary, Secondary, Advanced, NorthLive, Distance, CreditSources, SavingSources). The dependent variable, Member, is a dummy variable taking value 1 if the household is a member, and zero otherwise. Independent variables include a series of income quintile indicators, YQuint1 through YQuint4, that reflect each respondent's position in the local cash income[2] distribution for the previous year. We omit the lowest-income quintile, with the other indicator variables reflecting increased local income (YQuint4 is the top 20% of earners locally).

Asset quintile indicator variables, WQuint0 through WQuint4, were similarly created from the approximate value of each respondent's liquid assets (land, livestock, radios, televisions, watches and other consumer durables). As with the income quintiles, the lowest asset quintile is excluded from the regression. If FSAs indeed serve their target group of the poorest 20% of the population, the coefficient estimates associated with each of the income and wealth quintile variables should be negative.

[2] Cash income was computed as the sum of salaries, business income, agricultural sales, livestock sales and other cash sources.

Primary, Secondary and Advanced are dummy variables measuring the respondent's highest level of education completed. If those with higher levels of education typically have access to higher-paying jobs and better financial services (even if our coarse measures of asset wealth and income do not fully capture this), and if FSAs serve the poor and uneducated who lack good access to financial services, the coefficient estimates on these variables should be negative as the omitted category are those who have not completed primary schooling.

The dummy variable NorthLive equals 1 if and only if the respondent belongs to an FSA in Marsabit district and holds their assets exclusively in livestock; it equals zero otherwise. If those pastoralists in the north invested exclusively in livestock are managing risk by utilizing the investment and savings opportunities afforded them by the FSA, we should expect to find a positive estimated coefficient associated with the NorthLive variable. Several commentators (e.g. Desta, 1999) have called for the expansion of financial services options among these underserved populations, so we investigate whether FSAs facilitate risk diversification for pastoralists.

The independent variable Distance measures distance from the FSA to the respondent's home in minutes via their typical mode of transportation. This variable might have either a positive or a negative relationship with membership. Those located further from the FSA bear the additional cost of travel to and from the FSA. However, those further away from the central town or village location of the FSA might be in greater need of FSA services as availability of financial alternatives decreases with distance from town.

Does access to loan alternative financial services decrease the likelihood of joining an FSA? The variable CreditSources represents the total number of non-FSA (formal and informal) loan sources the respondent accessed in the last year. Typical sources for loans include family, friends, shopkeepers (in the form of goods), traditional banks, moneylenders, NGOs and ROSCAs. This variable takes on a zero value if the individual did not take a non-FSA loan in the last year. The availability of loan alternatives is hypothesized to have a negative impact upon the decision to become a member. Those with fewer loan opportunities are more likely to join the FSA in the attempt to avail themselves of emergency liquidity.

The count variable SavingSources similarly measures the number of places the respondent used in the last year to safeguard cash, excluding the FSA. Typical storage facilities include one's home, one's shop, with a shopkeeper, with traditional banks and with savings groups. Individuals whose marginal need for security is high might exhibit greater demand for FSA investment and savings. On the other hand, if the availability of other savings mechanisms crowds out the need for FSA services, the coefficient estimate might be negative.

We estimated a probit model to test these hypotheses, with robust standard errors clustered on the region to preserve asymptotic validity in the presence of potential heteroskedasticity (Table 8.4). Contrary to the intent of FSAs, the estimation results indicate that the likelihood of FSA membership increases with both yearly annual income and asset wealth. All of the higher-quintile coefficients exhibit a positive relationship with Member, with six out of eight statistically different from zero. Respondents who completed post-secondary levels of

**Table 8.4.** Probit results for determinants of FSA membership.

| Variable | Coefficient | Robust standard error | Marginal effect |
|---|---|---|---|
| YQuint1 | 0.012 | 0.098 | 0.005 |
| YQuint2 | 0.272*** | 0.088 | 0.104 |
| YQuint3 | 0.138*** | 0.025 | 0.054 |
| YQuint4 | 0.174*** | 0.036 | 0.067 |
| WQuint1 | 0.445** | 0.216 | 0.168 |
| WQuint2 | 0.072 | 0.159 | 0.028 |
| WQuint3 | 0.178*** | 0.017 | 0.069 |
| WQuint4 | 0.547*** | 0.105 | 0.204 |
| Primary | 0.066 | 0.183 | 0.026 |
| Secondary | –0.119 | 0.131 | –0.047 |
| Adv | –2.176*** | 0.737 | –0.594 |
| NorthLive | 0.024 | 0.065 | 0.010 |
| Distance | 0.002* | 0.001 | 0.001 |
| CreditSources | –0.183*** | 0.037 | –0.072 |
| SavingSources | 0.610*** | 0.053 | 0.239 |
| Constant | –0.955*** | 0.121 | |

$n = 574$; pseudo $r^2 = 0.2716$.

Notes: ***signifies estimates that are statistically different from zero at the 1% level; **at the 5% level; and *at the 10% level.

Since FSAs in the same regions tend to share characteristics, robust standard errors are used to take into account heteroskedasticity and inter-region correlation (White, 1980).

education are statistically significantly less likely to join the FSA, although this is a very small subsample (13) and the effects of primary and secondary education on membership are statistically insignificant once one controls for relative income and wealth. Overall, it appears that the relatively poor and uneducated are not more likely to become FSA members, indicating that KDA is not reaching its target population.

Distance from the FSA bolsters membership, as reflected in the positive estimated coefficient associated with Distance. Those who live 1 h from the FSA are an estimated 5% (60 minutes times a marginal effect of 0.00093% per minute) more likely to belong to the FSA than someone who lives next door. This likely reflects the fact that informal financial service alternatives, such as accounts with shopkeepers or relatively wealthy town-dwelling relatives, decrease with distance from centrally located FSAs, and demand for membership increases with diminishing alternative sources of financial services. If a respondent has access to loans, he or she has little incentive to join the FSA, as suggested by the negative and significant coefficient estimates on CreditSources. In contrast, SavingSources has a statistically significant and positive relationship with FSA, indicating that those who actively save are more likely to join the FSA.

## 8 Purchase of Multiple FSA Shares

To what extent do those who belong to the FSA utilize its services more or less intensively? We consider participation first in terms of shareholdings, both because greater share purchases signal heavier equity investment in the FSA and because of the critical relationship between shareholdings and loan size. Having chosen to join the FSA, what motivates some members to purchase additional shares? The number of shares one owns determines one's borrowing and saving limits with the FSA, so we would expect share ownership conditional on membership to be driven largely by demand for credit and savings products.

Before delving into the regression model, we note that of the 282 FSA members interviewed, 180 purchased only one share, the minimum required for FSA inclusion. Of the 102 members who purchased multiple shares, over half purchased either two or three shares. Amongst multiple shareholders, 46% do not save at the FSA, as opposed to 56% of single-share shareholders. It therefore seems implausible that members buy multiple shares so as to increase savings ceilings.

Table 8.5 presents a breakdown of loans accessed by members according to shareholding categories. Of the 180 members who own a single share, only 12% borrowed from the FSA. In contrast, of the 70 members who own between two and five shares, 36% borrowed from the FSA, as did 64% of the 48 members holding more than five shares. Members with large shareholdings not only borrow more frequently than do single-share holders, but they also borrow larger amounts. Seven members (2% of the sample) holding more than 10 shares each received 11 loans worth 26% of the total value of loans disbursed, more than the 240 (80.5%) members who owned three or fewer shares combined. Clearly, multiple-share purchases are strongly associated with highly concentrated lending patterns.

The following Poisson count data model estimates the relationship between number of FSA shares owned by the respondent and member characteristics.

NumberShares = *f*(YQuint1–4, WQuint1–4, Savings000, Livestock, North, NorthLive, Headcount, InverseMills), where NumberShares is the total

**Table 8.5.** Loans and loan value by shares held.

| Shares | Number of members | Members borrowing | Members borrowing (%) | Number of loans | Value of loans (KSh) | Value of loans (%) |
|---|---|---|---|---|---|---|
| 1 | 180 | 21 | 12 | 25 | 39,800 | 9 |
| 2 | 37 | 11 | 30 | 18 | 32,500 | 7 |
| 3 | 23 | 9 | 39 | 11 | 42,500 | 9 |
| 4 | 10 | 4 | 40 | 7 | 25,500 | 5 |
| 5 | 13 | 8 | 62 | 12 | 56,400 | 12 |
| 6 | 8 | 5 | 63 | 8 | 47,400 | 10 |
| 7 | 3 | 2 | 67 | 2 | 15,400 | 3 |
| 8 | 3 | 2 | 67 | 3 | 22,200 | 5 |
| 9 | 3 | 2 | 67 | 2 | 22,000 | 5 |
| 10 | 7 | 4 | 57 | 9 | 40,000 | 9 |
| 10+ | 11 | 7 | 64 | 11 | 123,000 | 26 |
| Total | 298 | 75 | 25 | 108 | 466,700 | 100 |

number of shares owned by a member respondent. We use the same income and asset distribution indicator variables described for the membership probit. Savings000 calculates the total amount of savings held by the respondent, including deposits held at FSAs, measured in thousands of KSh. If access to higher savings limits motivates the purchase of shares, we would expect a positive estimate for the Savings000 coefficient. On the other hand, if loan access is the primary impetus driving the purchase of multiple shares, significant cash savings may obviate members' need for a sizeable loan and reduce the incentive to purchase multiple shares.

Livestock is a binary variable taking on value 1 if the respondent has only livestock assets to liquidate in times of need, and zero otherwise. If the Livestock coefficient estimate is positive, we can conclude that members whose sole asset is livestock use the FSA as an opportunity to diversify risk, as hypothesized by Desta (1999). Livestock is positively correlated ($r = 0.47$) with the North dummy variable, which takes value 1 if the respondent is a member of one of the four FSAs located in Marsabit district. The interactive dummy NorthLive is the product of the North and Livestock variables. All three dummies are included to specifically test for multiple shareholdings amongst those transitioning out of pure pastoralism in the dry northern regions of Kenya. The variable North is expected to have a negative coefficient. Traditional dependence upon livestock, lower incomes and a conspicuous lack of profitable entrepreneurial opportunities render the requisite cash for share purchase a rarity. As we will see below, northern FSA members are less-active savers than are FSA members elsewhere. Lack of savings dampens one's incentive for purchasing multiple shares. The negative return to share investment renders shareholding an unprofitable investment, negating another motivation to purchase multiple shares. In the northern FSAs, the sole reason for purchasing multiple shares is to access larger loans.

A positive estimated coefficient of NorthLive would indicate that pastoralists use FSAs to diversify their asset risk out of livestock. The GL-CRSP PARIMA panel data reveal that the average December 2000 price of a male goat and a male sheep in North Horr was KSh 711 and KSh 870, respectively, equivalent to 2–3 FSA shares. Survey data reveal that the average North Horr FSA member's livestock holdings included 44 smallstock (sheep and goats), equivalent to at least 80 FSA shares. In addition, no member held fewer than 10 smallstock. If pastoralists prefer to diversify risk and invest or save in the FSA, the regression should yield positive coefficient estimates for NorthLive. Similarly, for members in the Coast or Nyanza provinces holding asset wealth exclusively in livestock, heavy investment in the FSA might be a good risk management strategy, so we predict a positive relationship between Livestock and FSA shares held.

We control for regional poverty by including the Headcount independent variable, as defined previously. Members from poorer locations with limited financial service availability might find FSA investment particularly attractive, leading to greater share purchases at the intensive margin.

Finally, we include the inverse Mills ratio, InverseMills, as calculated from the membership probit estimated in Section 7. If the coefficient estimate on the inverse

**Table 8.6.** Poisson regression results for number of shares owned.

| Variables | Coefficient | Standard error | Marginal effects |
|---|---|---|---|
| YQuint1 | 0.167 | 0.149 | 0.056 |
| YQuint2 | 0.148 | 0.141 | 0.049 |
| YQuint3 | 0.256** | 0.131 | 0.088 |
| YQuint4 | 0.697*** | 0.131 | 0.284 |
| WQuint1 | −0.412*** | 0.141 | −0.117 |
| WQuint2 | −0.427*** | 0.140 | −0.121 |
| WQuint3 | −0.061 | 0.131 | −0.019 |
| WQuint4 | −0.252** | 0.133 | −0.075 |
| Savings000 | −0.006*** | 0.002 | −0.002 |
| Livestock | −0.174* | 0.100 | −0.055 |
| North | −0.405** | 0.184 | −0.121 |
| Northlive | 0.124 | 0.225 | 0.041 |
| Headcount | −0.002 | 0.002 | −0.001 |
| Invmillsrat | −1.384*** | 0.107 | −0.442 |
| Constant | 1.466 | 0.179 | |

$n$ = 574; pseudo $r^2$ = 0.181.

Notes: ***signifies estimates that are statistically different from zero at the 1% level; **at the 5% level; and *at the 10% level.

Since FSAs in the same regions tend to share characteristics, robust standard errors are used to take into account heteroskedasticity and inter-region correlation (White, 1980).

Mill's ratio is significantly different than zero, this corrects for the selection bias associated with choosing to become a member of the FSA (Heckman, 1979).

Table 8.6 presents the results of the Poisson regression. If multiple share ownership were driven by demand for diversification into savings, the correlation coefficient with non-financial wealth would be positive. The fact that multiple FSA share ownership is instead negative and statistically significantly associated with household non-financial wealth underscores that multiple share ownership appears driven by members' demand for credit, as opposed to savings. This is also reflected in the strongly positive and statistically significant relation between household income and multiple share ownership. It takes money to buy shares and, as we shall see, to borrow money.

This inference is further reinforced by other regression coefficient estimates. The negative and statistically significant coefficient estimate on the inverse Mills ratio signals that holding other member characteristics constant, multiple share ownership is negatively associated with the probability of membership. This result suggests that membership is undertaken strategically by those seeking the relatively high levels of credit made accessible through multiple share ownership. Similarly, holding other cash savings crowds out multiple FSA share ownership, as indicated by the negative coefficient estimate associated with the Savings000 variable. Members holding their assets exclusively in the form of livestock as well as members of the northern FSAs are less – not more – likely to purchase multiple shares. Substantial decline in northern FSA share values renders multiple shareholdings unattractive, mirroring McPeak's (2005) observation that livestock remains a more remunerative form of savings than do financial assets held in banks.

# 9 Use of FSA Savings Instruments

As explained previously, the FSA generates its own capital entirely from sales of shares because the reserve requirement on savings is 100%. What explains members' decisions to avail themselves of the savings instruments offered by FSAs? Of the 299 members interviewed 34% had used FSA savings, though this ratio drops to less than 5% of the 88 northern FSA members. Members holding savings balances were asked about their motivations for savings with the FSA. Conversely, members who do not save at the FSA were asked why they choose not to.

The most frequently cited reason for saving with the FSA was easy access to deposits in times of need. In addition, savers appreciated FSA proximity, security and the opportunity to accumulate capital to achieve long-term investment goals. A few of the Kwale and Migori FSAs were in the process of requiring savings to access loans, as reflected in nearly 8% of savers citing the desire to access loans as the motivation for saving at the FSA.

Lack of funds dominated the reasons offered for not saving at the FSA. Lack of accessibility and lack of security ranked a distant second and third. All the respondents who did not save at the FSA for lack of trust in the institution were from the north, including the wife of the Chairman of the Board of one FSA. Similarly all who attribute not saving at the FSA to lack of information about savings services are from the north, as are 80% of those concerned with the zero interest paid to deposits. Despite the scarcity of savers at northern FSAs, there might exist a latent demand for savings not serviced by FSAs due to poor community image, lack of trust, stagnant rate of return and poor information dissemination.

# 10 Borrowing from the FSA

Of the sampled FSA membership, only 15% borrow from the FSA. The 85% investing in the FSA without borrowing bear, in the value of their shareholdings, the default risk of the minority of members who do take out loans. We already observed that the likelihood of borrowing generally increases in the number of shares owned, which is itself positively related to income and negatively associated with non-financial wealth, including non-diversified holding of livestock assets. We now explore in greater detail the question of which members borrow from the FSA.

We investigate this question using the following probit regression model with the dependent variable FSALoan, which equals 1 if the member borrowed from the FSA, and zero otherwise:

FSALoan = *f*(YQuint1–4, WQuint1–4, NumberShares, CreditSources, Savings000)

Most of the independent variables have been previously defined, particularly the income and wealth quintile variables. In order to further define the relationship between shareholdings and borrowing, the variable NumberShares is

**Table 8.7.** Probit regression results for FSA borrowing.

| Variable | Coefficient estimate | Standard error |
|---|---|---|
| YQuint1 | 0.485 | 0.312 |
| YQuint2 | 0.116 | 0.328 |
| YQuint3 | 0.134 | 0.325 |
| YQuint4 | 0.869*** | 0.331 |
| WQuint1 | –0.097 | 0.303 |
| WQuint2 | –0.655* | 0.336 |
| WQuint3 | –0.613* | 0.334 |
| WQuint4 | –0.521 | 0.345 |
| Numshares | 0.072*** | 0.019 |
| Creditsources | 0.024 | 0.090 |
| Savings000 | –0.112** | 0.055 |
| Constant | –0.914** | 0.361 |

$n = 489$; $\chi^2 = 32.44$ ($p$ value = 0.00).

Notes: ***signifies estimates that are statistically different from zero at the 1% level; **at the 5% level; and *at the 10% level.

Since FSAs in the same regions tend to share characteristics, robust standard errors are used to take into account heteroskedasticity and inter-region correlation (White, 1980).

included as a regressor. Controlling for the number of shares a member owns, we seek to establish whether borrowing is related to income or wealth, and to substantiate our earlier hypothesis that borrowing demand motivates the purchase of multiple shares, controlling for income and wealth.

Estimation results are presented in Table 8.7. The coefficient estimates associated with the income quintiles suggest that it is not the poorest 20% of the population that receives loans. Rather, members in uppermost income quintiles are the most likely to receive FSA loans, contrary to KDA's stated objective. Likelihood of borrowing is weakly, negatively related to non-financial wealth. The number of FSA shares owned has a strongly positive and statistically significant effect on borrowing, even controlling for member income and wealth, which we earlier established are key determinants of multiple share ownership. This reinforces our conclusion that multiple share ownership is motivated primarily by members' desire to access loan capital.

The negative coefficient estimate associated with the non-FSA savings variable indicates that those with adequate savings are less likely to borrow from the FSA. Savings provide a substitute vehicle for members to self-insure against income shocks and to accumulate investment capital. Since the opportunity cost of savings, current consumption, is often lower than the cost of interest-bearing loans, members with liquid savings are commonly better served saving rather than borrowing.

## 11 FSA Loan Repayment

We have established that higher-income individuals are more likely to become members of the FSA. Conditional upon having chosen to join the FSA, those members with the highest incomes are more likely to borrow and to purchase multiple shares, allowing them to access higher-value loans. We have also seen that FSA share values have been declining, on average, across KDA's FSAs, with the decline related to the number of generations of loans the FSA has made. This strongly indicates that loan repayment is a problem for FSAs. The borrowing pattern evident in the data, including multiple share ownership patterns, raises the possibility that FSAs are not serving the poor as delineated in KDA's stated goals. Given that poorer members are less likely to borrow but hold shares which erode at the same rate as all other members' shareholdings, and that higher-income members are more likely to borrow, and in larger amounts, FSAs might be inadvertently facilitating de facto transfers from poorer to wealthier members if these better-off borrowers are not reliably repaying loans. Exploring loan repayment patterns becomes essential if we are to understand the distributional effects of FSAs in Kenya.

Analysing loan repayment behaviour is tricky, however. Of the 298 members in the sample 75 took loans ranging from KSh 500 to KSh 35,000, with a mean of KSh 6223 and a median of KSh 3600. But when we asked about loan repayment behaviour, all but eight respondents claimed to have repaid their loans on time, and those eight reported suffering no negative consequences as a result of their delinquency. Evidence presented by FSA records disputes the veracity of these payment claims, as we demonstrate below. Reconstructing loan repayments from respondent recall is unconstructive, so we turn from the survey data to an alternate data source for loan repayment analysis: FSA records.

The quality of FSA records fluctuated from site to site, ranging from well-kept general ledgers to scraps of cash-in and cash-out vouchers. FSAs varied tremendously in their thoroughness of bookkeeping due to managers' education level, training received from KDA, frequency of FC contact, staff turnover rate as well as the detail of information demanded by the local Board of Directors. It was not uncommon for the staff and board members to have no idea about FSA share value, or how to compute it. Share value was often calculated exclusively by KDA auditors who arrived once each year to evaluate books, confirm deposits and assess share value.

We gathered data on 894 loans, or 91.6% of the 983 total loans issued by nine FSAs. Loans from the remaining five sampled FSAs were not included in the analysis due to lack of data either owing to time constraints (Rongo, Magunga and Msambweni), because loans had not yet been issued (Kikoneni) or because loans issued were not yet due (Kalacha). Although 983 loans were issued by these nine FSAs, we were unable to extract information from 89 of them due to missing records. This near-census of loans was taken out by a broader range of members than those randomly sampled. Unfortunately, these data lack some of the information the survey included, most notably member income, asset and characteristic information. We are not aware of other published studies that analyse microfinance data at this level of detail.

The 894 loans under examination can be divided into two broad categories: regular and emergency. Regular loans require approval of the FSA-specific Credit Committee and can take up to a full month before approval and disbursement. The regular loan application is quite cumbersome, requiring information such as years at current residence, loan purpose, sources of income, projected income and the signatures of two guarantors. In addition, the FSA manager must assess the moral character of the applicant and recommend that the Credit Committee review the loan application. The terms of regular loans vary from FSA to FSA, but generally are to be repaid monthly over a 3- to 6-month term at an interest rate of 10% per month on a reducing balance. Of the loans under investigation 83% are regular loans.

In contrast, emergency loans are issued upon demand at the manager's discretion. The cost of the expedited process is reflected in the shortened, 1-month term, as well as a higher interest rate, between 12% and 15% per month. The significance of emergency loans varies widely across FSAs. While over one-third (52 of 145) of Karungu's loans were emergency, at the other extreme, Tiwi and Kalacha did not issue emergency loans at all. Emergency loans are generally smaller than regular loans, and comprise only 6.5% of total loan value, although they represent 17% of all loans issued.

The distinction between emergency and regular loans reflects different screening mechanisms. Emergency loan evaluation is the exclusive domain of managers. As opposed to regular loans that complement manager evaluation with peer screening by the FSA Credit Committee, emergency loans rely on peer monitoring solely to guard against moral hazard or loan default *ex post* of a lending decision. If *ex ante* peer screening reduces the lending risk associated with adverse selection, this should render emergency loans riskier than regular loans; hence the different terms on the loan types. Comparison of emergency and regular loan repayment performance, while controlling for loan size, permits us to separate the effects of *ex post* peer monitoring (present in both) from *ex ante* peer screening (present in regular loans only).

## 11.1 Loan repayment

Of the 894 FSA loans issued in the sample, 59% (579) were paid in entirety, 13% (113) received no principal repayments and principal was partially repaid for the remaining 28% (202). Thus, 35% (315) of the 894 loans were in arrears, defined as loans with principal past due. The amount in arrears represents 24% of the total value of loans issued, and 64% of total share capital. On average, 61% of principal due had been paid on the 315 loans in arrears.

Of the loans in arrears, 13% are emergency loans, which represent 17% of all loans. Although we hypothesized that the additional rigour of the regular loan approval process might cause emergency loans to default at a higher rate than regular loans, this does not appear true, implying that the Credit Committee does not effectively screen out high-risk loans. Indeed, scant discrimination among loan applications is evident. For example, the Credit Committee turned down only 2.4% (3:340) of loan applications to the North Horr FSA. This calls into question the assumption that the local Board of Directors harnesses superior local knowledge so as to protect FSA assets against loans with high levels of *ex ante* credit risk. Mude (forthcoming) posits that the intricate nature of social relations binding community members makes it personally beneficial for volunteer Board of Directors members to issue loans, despite a priori shared knowledge that applicants are bad credit risks. Rejecting loan applications *ex ante* or enforcing penalties *ex post* results in disutility borne personally by FSA officials in the form of rancor, ill-will, strained social relations, etc. On the other hand, the disutility of delinquent loans is distributed across all shareholders. If the personal costs to officials of loan applicant rejection outweigh the personal costs these officials bear as shareholders due to issuing risky loans, it becomes individually rational for FSA officials to issue loans expected to underperform, and not to pursue non-performing loans. Of course, this 'loans makes friends' hypothesis turns the canonical logic of Stiglitz (1990) – that 'friends make loans' – on its head.

It is informative to look at loan repayments rates by loan size. Recall that the upper bound on loan size is a direct function of shares held by the borrower, which we earlier established is strongly and positively associated with income. If the relatively poor face more difficulties in repaying their loans, we would therefore expect repayment rates to vary inversely with loan size. However, descriptive statistics reveal that loans of KSh 1200 and below have a rate of default (39%) no higher than any other loan group. In fact, the largest loans, of KSh 15,000 and more, suffer the highest default rate (53%). Although shareholding is positively associated with income, it does not appear to signal increased propensity to repay.

To explore repayment rates in more detail, we estimate the following doubly censored Tobit model, censored at both zero and 100%, relating loan repayment percentage to loan size, type of loan and location as captured through site-specific dummy variables.

%Repaid = *f*(Principal000, Emergency, Headcount, AvgAltern, SiteSpecific Dummies) where %Repaid is a continuous variable from zero to 1 capturing the percentage of principal repaid on a loan. Principal000 reflects loan size (in thousands of KSh). If larger loans are indeed more prone to default, we would expect a negative coefficient estimate on Principal000. Emergency is a dummy variable, equaling 1 if the loan issued was an emergency loan, and zero for a regular loan. Although theory predicts that less carefully screened emergency loan applications should default at a higher rate, the descriptive statistics suggest no such unconditional effect. Moreover, since emergency loans are smaller and the unconditional descriptive statistics suggest that larger loans default at a higher rate, the loan size effect could be masking the loan screening effect on repayment performance.

Headcount once again measures the local incidence of poverty. If members from poorer areas have additional incentives to maintain their access to FSA financial services, we would anticipate a positive relationship between the percentage of the population living beneath the poverty line and the percentage of loan principal repaid. AvgAltern measures the site-specific average number of non-FSA credit and savings alternatives used by respondents within the last year, as extrapolated from the 574 individual surveys referenced earlier. We would predict that the more financial service alternatives available to, and used by, respondents, the less likely respondents are to safeguard their reputation with the FSA and the lower their probability of repaying principal owed.

**Table 8.8.** Tobit regression results for proportion of principal repaid.

| Variable | Coefficient estimate | Standard error |
|---|---|---|
| Principal000 | −0.017** | 0.008 |
| Emergency | 0.064 | 0.117 |
| Headcount | 0.013*** | 0.005 |
| AvgAltern | −0.215** | 0.106 |
| LungaLunga | 0.691*** | 0.231 |
| Tiwi | 0.852*** | 0.201 |
| Karungu | 0.564** | 0.259 |
| BadhaHuri | 0.786*** | 0.255 |
| Korr | 0.565** | 0.253 |
| Constant | 0.178 | 0.444 |

$n = 894$; pseudo $r^2 = 0.164$.
Notes: ***signifies estimates that are statistically different from zero at the 1% level; **at the 5% level; and *at the 10% level.
Since FSAs in the same regions tend to share characteristics, robust standard errors are used to take into account heteroskedasticity and inter-region correlation (White, 1980).

Site-specific dummy variables control for characteristics particular to each FSA. Eight site-specific dummies each take on value 1 if the FSA is located at that site, and zero otherwise, with North Horr as the omitted site.[3]

The estimated relationship between principal repaid and repayment percentage is negative and statistically significant, confirming that repayment is decreasing in loan size (Table 8.8). Those with the largest loans default most. We already established that the probability of borrowing and the number of shares held (and thus a member's borrowing limit) are positively and significantly related to household income. Thus, it appears that the highest-income FSA members are most likely to default and the lowest-income members are least likely to borrow or default. The implication is that poorly performing FSAs provide a de facto mechanism for regressive transfers from lower-income non-borrowing members to higher-income, borrowing-and-defaulting members.

The coefficient associated with the dummy emergency variable, though positive, is not significant, implying that the distinction between emergency and regular loans is irrelevant in terms of loan performance and contravenes the standard theoretical prediction. Rather, it is consistent with Mude's (forthcoming) model of socially constrained lending behaviour. The one-day, one-person screening for emergency loans is just as ineffective in terms of screening out bad credit risks as the lengthy process involving the Board of Directors. Given equally effective *ex ante* screening and *ex post* monitoring, lower costs of administration, higher rates of interest and faster loan cycles, one wonders why other FSAs have not followed Karungu's lead, disbursing emergency loans with gusto.

[3] The other site dummy variables were ultimately dropped due to perfect multicollinearity with the other covariates.

The coefficient estimate on the Headcount variable is positive and significant, affirming that loans made in poorer areas exhibit higher rates of repayment. Note that this effect is independent of the relationship between poverty and the availability of alternative financial service providers, which has a negative and significant association with the percentage of principal repaid.

## 11.2 Implications of loan size and loan delinquency

The repayment rate difference between large and small loans has acute repercussions. The 269 loans of KSh 1200 or less represent a mere 7% of total portfolio value. In contrast, the 200 loans greater than or equal to KSh 10,000 comprised 60% of the principal borrowed (Table 8.9). FSAs are heavily exposed to default risk associated with large loans, which are typically made to higher-income members. Indeed, the value of the principal in arrears on the 20 loans greater than KSh 15,000 that have not fully repaid (KSh 286,006) is comparable to the total principal paid out in the smallest 269 loans (KSh 304,450). The 20 largest loans in default represent 14% of total share value for the total 2886 KDA members across nine sites. The minimum investment value of the shares required to access these large loans is greater than the total annual income of 66% of the survey respondents who purchased only one share.

Perhaps surprisingly, these figures give an optimistic view of the situation. Whether due to unwillingness or lack of training, most FSA managers do not calculate the interest and penalties on loans according to the standardized KDA accounting formulae. Using the original loan information, we revised the arrears calculations to conform to the FSA's official, published accounting terms: the regular interest rate of 10% per month paid on a reducing balance increases to 15% per month if principal is past due. In addition, penalties not

**Table 8.9.** Distribution of loans and official account of loans in arrears by loan size.

| Loan size (KSh) | 1,200 | 2,400 | 3,600 | 6,000 | 10,000 | 15,000 | 15,000+ |
|---|---|---|---|---|---|---|---|
| Number of loans | 269 | 134 | 114 | 178 | 100 | 53 | 53 |
| Loan value (KSh) | 306,450 | 293,150 | 374,000 | 974,300 | 860,700 | 649,600 | 1,177,600 |
| % of loan value | 7 | 6 | 8 | 21 | 19 | 14 | 25 |
| Loans in arrears | 104 | 57 | 39 | 79 | 44 | 28 | 20 |
| Value in arrears | 79,265 | 79,054 | 88,814 | 263,359 | 249,792 | 230,676 | 286,006 |
| Value in arrears/ total value (%) | 26 | 27 | 24 | 27 | 29 | 36 | 24 |
| Value in arrears/ total arrears (%) | 6 | 6 | 7 | 21 | 20 | 18 | 22 |

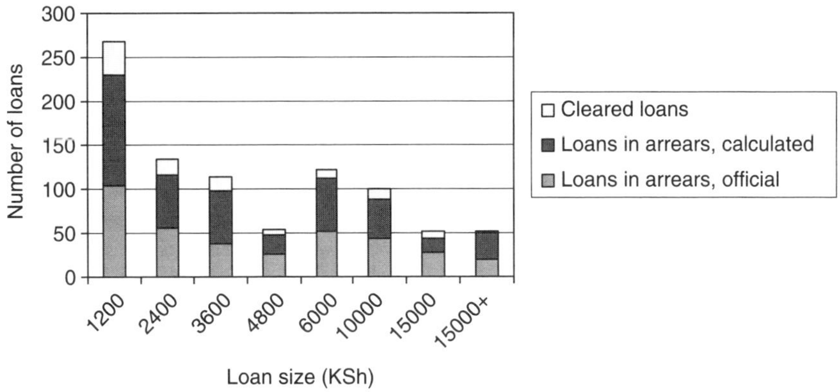


**Fig. 8.1.** Distribution of loans across loan size, correctly calculated.

collected are compounded with the principal outstanding, and themselves are penalized at the rate of 15% per month. Late payments made on the revised accounts were applied against interest and penalties accrued before principal.

With these accounting corrections, the performance of FSAs worsens. After recalculation, 729 of the total 894 loans were in arrears, totaling 39% of loan principal, or 78% of share capital. In contrast, the official accounts calculated principal outstanding as 24% and 64% of loan principal and share capital, respectively. Accurate accounting more than doubles the number of loans with 90–100% of principal past due from 88 to 190. When we look at the distribution of the numbers of loans and arrears across loan size (Figs 8.1 and 8.2), it becomes evident that the vast majority of loans are not repaid in their entirety. If penalties were properly assessed, few loans would show completely clear accounts. The value of the principal past due on the largest 206 loans of KSh 10,000 or more now increases to 25% of the total principal issued, or 53% of the current portfolio. The increased principal in default results in average share value loss of 78%, with over half of this loss attributable to 184 defaulting members, each holding loans of KSh 6000 or more.

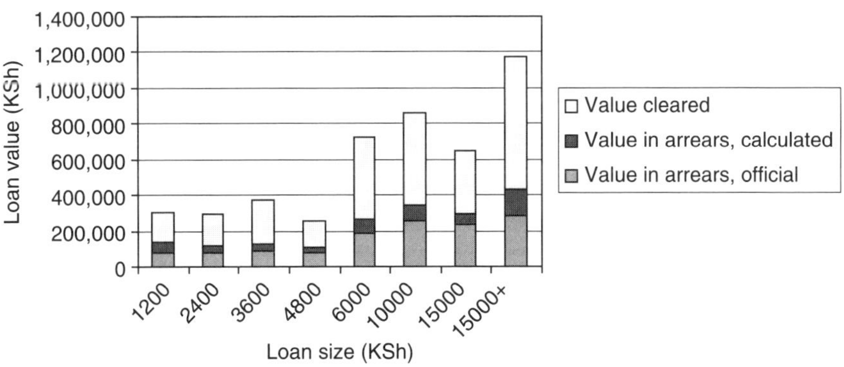


**Fig. 8.2.** Distribution of loan value across loan size, correctly calculated.

## 12 Conclusions

KDA FSAs serve primarily higher-income individuals, who not only are the most likely to become members, but are also most likely to borrow, to purchase multiple shares and thereby to take out larger loans, which we have demonstrated default at a higher rate than smaller loans. Unsustainably high rates of loan delinquency seriously erode share value and threaten FSA financial viability. Despite their stated mission, KDA FSAs typically bypass the poorest members of the communities they serve and implicitly fostering regressive wealth distribution from lower-income, non-borrowing members to higher-income members more likely to take out and default on large loans. An important lesson to learn is that without careful analysis of the patterns of benefits reaped and the bearers of costs borne, we cannot accept at face value that an initiative is 'pro-poor'.

So what can be done to improve this situation? First, by linking loan limits to shareholdings, FSAs reserve large loans for those with sufficient disposable income to purchase multiple shares. Decoupling loan size and shareholdings would allow for progressive lending, whereby larger loans are issued based upon an established credit history of successfully paying back smaller loans. In this manner, the FSA could better direct loans towards the creditworthy poor, as opposed to the relatively wealthy who can afford multiple shares. This policy change could have particular impact in the northern FSAs, where the loan size available to poorer members holding only one share is insufficient to cover the costs of transportation into town, thereby effectively precluding productive investment of borrowed capital.

Improved *ex ante* screening and *ex post* monitoring of loans is plainly necessary. Protection of equity investment does not appear to provide adequate motivation for rigorous loan application screening and loan monitoring by boards of directors. Although this decentralized approach taps into local knowledge, it can also be encumbered by local-level power relations and social considerations that make it rational for the Board to approve loans to uncreditworthy borrowers and not to pursue bad debt (Mude, forthcoming). It might be worth exploring compensation schemes for Board members who supervise a healthy portfolio so as to offset the personal costs incurred when boards reject loan applications and exercise personal influence and moral suasion to recover delinquent loans.

To protect share capital, additional training of staff and board members in loan screening, accounting and management is required to confer basic tools for managing an MFI. In the FSA model, joint liability reigns in the entire membership as all members have an equity investment to protect. However, if the bookkeeping is not accurate, if a list of defaulters is not compiled, if current share value is an unknown, if vital statistics about the health of the FSA is not available, members are not empowered with the requisite knowledge necessary to protect their MFI. Lack of technical skills including bookkeeping, interest and penalty calculation, auditing and computation of current share value threatens the FSA.

To avoid the inefficiencies of multilayered bureaucracy, funds were designated by KDA's donors for the opening of FSAs as opposed to the maintenance of FSAs. Limiting investment to start-up costs, however, results in insufficient administrative support for KDA's support operations, including training of FSA staff, FSA board members, and KDA FCs. Particularly in the north, with poor roads and infrequent passenger vehicles, FSAs operate almost independently, with negligible help from FCs who have vast territories, no vehicles and limited institutional support.

A shift in priorities from loan issuance to providing secure savings might be the most effective means to improve FSA financial viability and to serve the poor. This would require reducing fixed fees (e.g. passbook charges) and offering interest payments so that smaller deposit volumes can earn a reasonable rate of return. It would also require relaxing the present 100% reserve requirement on savings deposits. Without a structural change in FSA by-laws, however, FSAs will continue to prefer to collect share capital to invest rather than to amass unproductive savings deposits.

## Acknowledgements

We thank Miriam Cherogony and K-REP Development Agency (KDA), who provided invaluable guidance and unfettered access to KDA data. This work has been made possible by support from the United States Agency for International Development (USAID), through grants LAG-A-00-96-90016-00 to the BASIS CRSP, DAN-1328-G-00-0046-00 and PCE-G-98-00036-00 to the Pastoral Risk Management (PARIMA) project of the Global Livestock CRSP, and the Strategies and Analyses for Growth and Access (SAGA) cooperative agreement, number HFM-A-00-01-00132-00, as well as the Mario Einaudi Center for International Studies at Cornell University. Special thanks to John and Margaret McPeak for their comments and hospitality, and to Ralph Christy, Andrew Mude and John Omiti for helpful comments and discussions. Any remaining errors are ours alone.

## References

Adams, D.W. and von Pischke, J.D. (1992) Microenterprise credit programs: déjà vu. *World Development* 20(10), 1463–1471.

Besley, T. and Coate, S. (1995) Group lending, repayment incentives and social collateral. *Journal of Development Economics* 46(1), 1–18.

CBS (Central Bureau of Statistics) (2001) *1999 Population and Housing Census*. Vol 1: *Population Distribution by Administrative Areas and Urban Centers*. CBS, Ministry of Planning and National Development, Nairobi, Kenya.

CBS (Central Bureau of Statistics) (2003) *Geographic Dimensions of Well-Being in Kenya: Where Are the Poor?* Vol 1: *From Districts to Locations*. CBS, Ministry of Planning and National Development, Nairobi, Kenya.

Christen, R.P. (1998) Keys to financial sustainability. In: Kimenyi, M.S. Wieland, R.C. and von Pischke, J.D. (eds) *Strategic*

*Issues in Microfinance*. Ipswich Book, Suffolk, England, pp. 193–186.

Conning, J. (1999) Outreach, sustainability and leverage in monitored and peer-monitored lending. *Journal of Development Economics* 60(1), 51–77.

Desta, S. (1999) Diversification of livestock assets for risk management in the Borana pastoral system of southern Ethiopia. PhD dissertation. Department of Rangeland Resources, Utah State University, Logan, Utah.

Government of Kenya (2000) *Incidence and Depth of Poverty*, Vol. 1. Ministry of Finance and Planning, Nairobi, Kenya.

Heckman, J. (1979) Sample selection bias as a specification error. *Econometric* 47(1), 153–161.

Hollis, A. and Sweetman, A. (1998) Microcredit: what can we learn from the past? *World Development* 26(10), 1875–1890.

Hulme, D. and Mosley, P. (1996) *Finance Against Poverty*. Routledge, London.

Jazayeri, A. (1996) Financial Services Association (FSA) – Concept and Implementation Arrangements, Document. IFAD.

McPeak, J.G. (2005) Individual rationality and collective rationality in pastoral production: evidence from Northern Kenya. *Human Ecology* 33(2), 171–197.

Morduch, J. (2000) The microfinance schism. *World Development* 28(4), 617–629.

Mude, A. (Forthcoming) Making loans to make friends: explaining the dismal performance of financial service associations. *Agricultural Finance Review*.

Pearce, D. and Helms, B. (2001) Financial service associations: the story so far. Document. Consultative Group to Assist the Poorest.

Pederson, G.D. and Kiiru, W.K. (1997) *Kenya Rural Enterprise Program: Case Study of a Microfinance Scheme*. World Bank, Washington, DC.

Stiglitz, J.E. (1990) Peer monitoring and credit markets. *World Bank Economic Review* 4(3), 351–366.

White, H. (1980) A heteroskedasticity-consistent covariance matrix estimator and a direst test for heteroskedasticity. *Econometrica* 48(4), 817–830.

Woolcock, M.J.V. (1999) Learning from failures in microfinance: what unsuccessful cases tell us about how group-based programs work. *The American Journal of Economics and Sociology* 58(1), 17–42.

Yaron, J. (1994) What makes rural finance institutions successful? *The World Bank Research Observer* 9(1), 49–58.

Zeller, M., Schrieder, G., von Braun, J. and Heidhues, F. (1997) *Rural Finance for Food Security for the Poor: Implications for Research and Policy*. IFPRI, Washington, DC.

# 9 The Effects of Farmer Community Group Participation on Rural Livelihood in Kenya

D.M. Amudavi

## 1 Introduction

Group participation is perceived to impact economic, social and political outcomes. Individuals' involvement in specific informal or formal organizations for the purposes of advancing self-interests and collective interests is an intrinsic aspect of human life and, depending on the nature and quality of the groups, may influence members' well-being (Stewart, 2005). A significant conceptual and empirical research agenda has risen from issues surrounding group participation (Ostrom, 1990; Uphoff *et al.*, 1998). Several of the debates focus on the efficacy of rural groups and their role as instruments for improving individual welfare (Craig and Mayo, 1995; Berner and Phillips, 2005; Robeyns, 2005; Stewart, 2005). However, there is little solid evidence to support this claim. This chapter examines whether group participation in rural Kenya substantially influences household welfare and whether this varies by group type and by the extent to which the groups enable farmers to access services and resources.

Groups may be important in promoting economic well-being and in buffering against natural and policy shocks and may affect peoples' sense of well being (e.g. through self respect) (Platteau and Gaspart, 2003; Grootaert and Narayan, 2004). Groups may be important instrumentally in determining efficiency and resource distribution, and in influencing peoples' values and choices (Stewart, 2005). Groups can facilitate low-cost access to information, thereby stimulating technology adoption and enhancing contract enforcement (Narayan and Pritchett, 1999; Knox *et al.*, 2002; Grootaert, 2004; Place *et al.*, 2004). They may facilitate labour sharing at critical times (Cleaver and Schreiber, 1994). Other important functions performed by groups include organizing market access, input supply, savings and credit, and informal insurance (Narayan and Pritchett, 2000; Weinberger and Jüting, 2000; Mknelly and Kevane, 2002). Based on economic functions, Heyer *et al.* (2002) divide

 *Decentralization and the Social Economics of Development* (eds C.B. Barrett, A.G. Mude and J.M. Omiti)

functions of groups into three types: *efficiency* functions, aimed at overcoming market failures; *claims* groups, intended to improve the share of resources or power of their members; and *pro bono* groups, typically aimed at providing benefits for others (usually among the poor) in society. The cited literature suggests that groups may matter to rural welfare.

There is, however, no reason to believe that all groups are equally beneficial to members. The ability of a group to generate any of these instrumental benefits for members may depend on social structures internal to the group, structures that determine the formulation and enforcement of rules, and define the manner in which collective decisions and actions are made and implemented (Katz, 1981; Serageldin and Grootaert, 2000). Group benefits also depend on their relationship with external agencies and organizations (Poole *et al.*, 1996; Krishna, 2002, 2003), and on the range of services they provide their members (Portes, 1998; Prokopy, 2005). Hence, differentiating between effective and ineffective group types and services coverage may be important, especially if rural development policy is aimed at stimulating the effective use of group-based services to advance poverty reduction objectives.

Despite the widespread current fascination with rural groups, there has been scant careful, empirical examination of the potentially differentiated effects of group participation by group type and by the benefits leveraged. Some relevant studies have focused on structural effects internal to the groups, such as effects of group heterogeneity and degree of member involvement in decision-making processes, but less on structural effects external to the functioning of groups (Grootaert, 2001; Grootaert and Narayan, 2004). Thus, analyses of group participation have largely ignored examining benefits derived from participation (Grootaert, 2004; Okten and Osili, 2004 are notable exceptions). These limitations hinder useful inference. Thus, understanding whether some groups matter more than others in influencing household welfare, and what makes them function differently from others, is necessary.

This chapter, therefore, explores the relationship between group participation and household welfare through empirical analysis of four propositions. First, there are differences in types of group participation across households contingent upon members' ability to meet membership requirements. This is a standard proposition tested in the empirical literature on groups and social capital (Gross *et al.*, 2001). Second, the level of differential group participation, as measured by the density or number of particular mediating structures, exerts an important and significant effect on welfare, beyond that accounted for by household resources. This distinguishes what type of group participation makes a difference, if participation by a critical mass of households matters or only participation by some in the village. Third, services and resources (benefits) accessed via different sorts of groups also exert an important and significant effect on welfare, beyond that accounted for by both household resources and number of group memberships. As observed by Narayan and Pritchett (2000), such benefits help overcome multiple government and market failures for group members. Lastly, the determinants of participation in distinct types of groups differ, with less universal access to groups offering greater benefits.

Answers to these propositions can inform policy analysis and design geared towards enhancing capacities of groups to support smallholders' livelihood.

Section 2 briefly reviews the social capital literature that relates welfare to group participation and group-based rural development programming in Kenya. Section 3 describes the methods and variables used in the analysis and presents descriptive statistics on these variables, on group participation patterns, trends in group participation, reasons advanced for both high and low group participation and benefits of group participation. Section 4 explores the effects of participation factors on household welfare measures, followed by a discussion of the determinants of participation and draws policy implications. Section 5 concludes by drawing attention to the potential of the two distinct types of rural groups in leveraging benefits and provides recommendations.

## 2 Social Capital, Rural Livelihoods and Development Programming in Kenya

Social capital is widely believed to facilitate vertical links between a community and outside social actors, making it a primary driver in maximizing access to services crucial to socio-economic welfare (Uphoff *et al.*, 1998; Lyon, 2003). According to Putnam (1995), social capital refers to features of social organization such as networks, norms and social trust that facilitate coordination for cooperation. Leana and Van Buren (1999) thus consider social capital to be both a private good, accruable to an individual, and a public good, one that benefits a society. Such capital originates in membership in voluntary associations, and in organizations such as community or local groups, which provide settings for people to learn to trust each other (Lyon, 2000). Portes (1998) defines social capital as 'the ability of actors to secure benefits by virtue of their membership in social networks and other social structures'. This, however, implies that it is not only the mere membership in such structures that matters, but also the emergent properties – services and functions – that address the intent of membership and the benefits associated with group participation. Widespread enthusiasm for the apparent benefits of social capital originating from associational life has understandably prompted interest in the use of local groups to promote rural development in Kenya.

Since Kenya attained independence from British rule in 1963, groups in Kenya have often been formed in the spirit of a mobilizing and organizing concept called *harambee* (Karega, 1996). The concept literally means social actors (individuals, groups or organizations), coming together to pool resources – human, financial, social and political – to address specific issues (Wallace and Wolf, 1999). Initially centred on sharing labour, the practice quickly spread to important areas such as education, health, conservation and politics. The achievements of some rural groups (e.g. self-help groups) captured public attention towards the potential of such groups to contribute to better rural livelihood (Karega, 1996; Gautam, 2000).

Underscoring the importance of community groups, Byrnes (2001) suggests that extension can most effectively carry out its mandate, not by working directly with individuals but by working with farmer groups and organizations.

A key role of rural extension is to facilitate increased agricultural production and rural development with the ultimate objective of improving welfare among rural households (Scarborough *et al.*, 1997; Wanga, 1999). This is achievable through bringing research-based knowledge in a usable form to groups, families and individuals in local communities (Hill and Parker, 2005). However, in Kenya, dwindling financial resources compounded by the effects of economic and institutional reforms associated with structural adjustment programmes have strained the subsistence economy and led to the collapse of requisite infrastructure for agricultural production (Rees *et al.*, 2000). Smallholders still require assistance to make good choices by expanding sources of knowledge and information, and enhancing their information processing capacity (Christoplos *et al.*, 2001). The National Agriculture and Livestock Extension Programme (NALEP) has restructured the national extension system through an increased focus on the private sector and other stakeholders to provide extension services (Republic of Kenya, 2001). The programme emphasizes the use of rural groups as an alternative pathway of addressing the complex and systemic issues facing smallholders (Wanga, 1999; Amudavi, 2003). The framework aims to improve on institutional responsiveness and accountability for sustainable service delivery (Antholt, 1994).

Over the last four decades, there has been dramatic growth in various types and configurations of rural groups. At a reasonably coarse level, one can distinguish between community groups that emerge endogenously within communities and supra groups that emerge at least partly at the instigation of, or with the intent of, linking to external agencies. I hypothesize that these two different types of groups might serve different functions for members and thereby have different effects on members' well-being.

Community groups are formed on locals' own terms and activities undertaken are self-identified and driven by members' own initiatives. Most community groups prefer to have identification; hence, they seek registration with the Department of Culture and Social Services, which requires groups to have at least 15 members. These groups are characterized by face-to-face interactions and personal participation is an important requirement. Community groups tend to enlist and engender the support and trust of a community, and are thus likely to succeed in attracting memberships (Zeller and Sharma, 1998). This is partly due to what McIntosh (2001) calls a 'problem-anchored helping network', in which members turn to others within their system of reciprocal exchange for assistance or advice.

Supra groups are formed in response to external development initiatives and have much larger memberships. They are formed by rural communities partly in collaboration with outside agencies usually in response to some anticipated resources, meaning that such groups often become vertically linked to outside institutions (Stringfellow *et al.*, 1997; Udvardy, 1998). They focus on specific objectives such as production, processing and marketing.

Building on Grootaert's (2004) initial assessment of local-level institutions to examine how community groups influence household welfare and poverty reduction, this chapter restricts its analysis to community and supra groups and the benefits they generate. The research is important because Kenya, like other developing countries, faces a considerable challenge in stimulating broad-based economic growth and effectively engaging with the global economy. Thus, any possible substantial outcomes provided to rural households through group participation offer economies of scope in the resolution of multiple government and market failures.

Undoubtedly, there are challenges in isolating the effects of differential group participation and outcomes associated with groups on welfare. However, empirical and theoretical work into the concept of social capital manifest in households' group participation suggests that, in the face of similar economic challenges, those households with high levels of group involvement will likely fare better than comparable households with lower levels of social capital (Trigilia, 2001; Prokopy, 2005). In turn, notable benefits from group participation may stimulate increasing demand for forming and joining of groups. This, however, will be contingent upon the efficacy of the groups to leverage the benefits (Akerlof and Kranton, 2000) and will reflect the group capabilities as collective entities that influence well-being (Stewart, 2005). This would suggest that an inability to either secure such access and rights, due to an individual's or group's inability, may undermine the sustainability of livelihood supported through collective entities (Cleaver, 2005).

## 3 Methods and Data Description

### 3.1 Study sites

This study was conducted between July 2003 and May 2004 in three districts in Kenya: Embu, and sample divisions Vihiga and Baringo (Fig. 9.1).

Embu district and sample divisions are close to Nairobi, the largest market in East Africa and located within eastern highlands (Republic of Kenya, 1997a). Its annual rainfall of 640–2000 mm occurs in March–June (long rains) and October–December (short rains). Soils are deep and of moderate to high fertility. Vihiga district and sample divisions are closer to Kisumu, the third largest city, located in the highlands of western Kenya (Republic of Kenya, 1997b). It receives an annual rainfall of 1800–2000 mm in April–June (long) and in September–November (short). It has well-drained dark red friable soils. On the other hand, Baringo district and sample divisions are distant from either city and situated within the Great Rift Valley (Republic of Kenya, 1997c). It is arid and semiarid, with an annual rainfall of 300–1200 mm in March–July (long) and September–November (short). It has well-drained soils on its highlands and complex alluvial soils in the lowlands.

### 3.2 Data collection and measures of group participation

Data from the three districts were collected in two phases. Phase one involved getting an inventory of agencies working with rural people including the public sector, non-governmental organizations (NGOs) and financial institutions; it also included an inventory of the rural groups. Phase two involved face-to-face

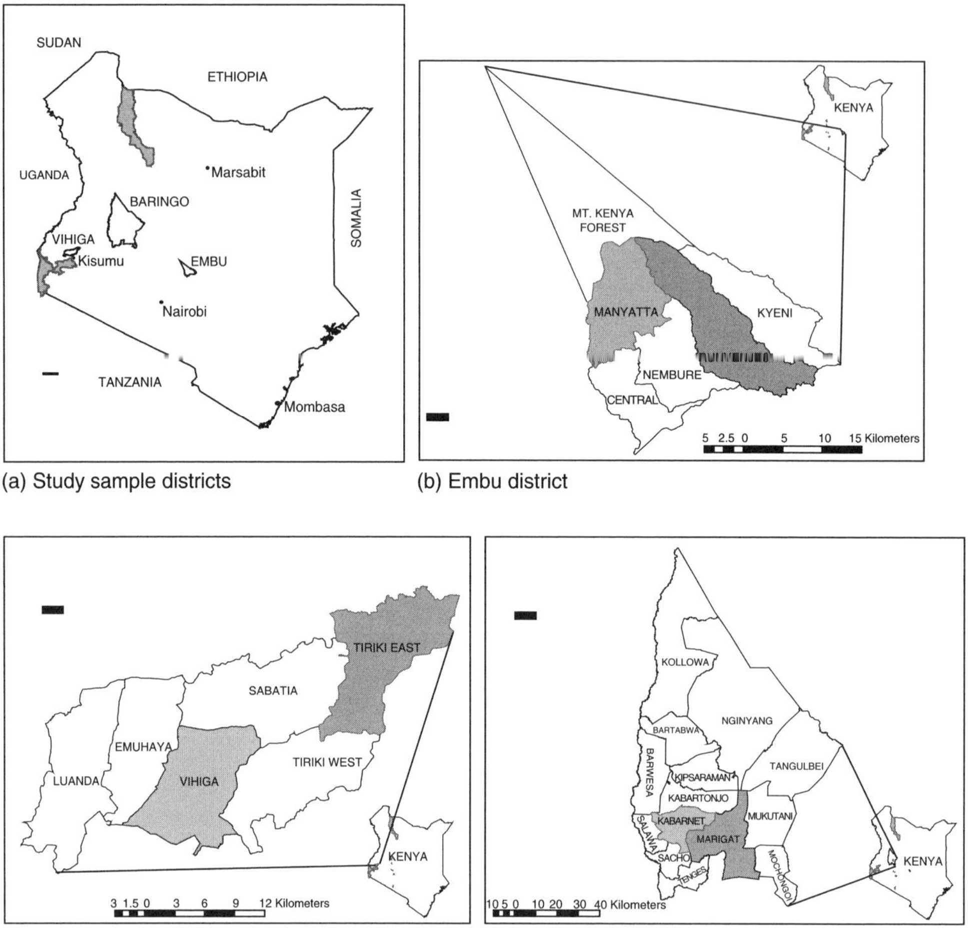


(a) Study sample districts (b) Embu district

(c) Vihiga district (d) Baringo district

**Fig. 9.1.** Study sample districts and sampling divisions (shaded).

interviews with 480 households selected using a combination of multistage, purposive sampling with simple random sampling. Six sub-locations (villages) were purposively chosen from NALEP focal areas and six from non-focal areas in two divisions from each district. A sample of 40 households was randomly selected from each village. Ten research assistants from the local areas were hired and trained to collect the data.

Household-level data were collected on participation in rural groups and on issues of collective action and trust. We held focus group discussions with 56 community groups on issues such as group formation and group size, group orientation, functions and benefits. The groups were identified by the extension staff working with farmers in the selected villages. All the sample groups were approached for a focus group discussion. The group participation measures included:

- *Types of rural groups.* Respondent's level of group participation was considered in two types of groups: (i) community groups formed endogenously by some people within a locality of their own accord based on their own needs, values and priorities; and (ii) supra groups formed exogenously by, or in cooperation with, external agencies (e.g. government, NGOs, private businesses) in response to some anticipated resource flow. Groups listed by a household were classified into either type.
- *Density of group participation.* This refers to memberships in either group type measured by the number of groups to which a household belonged. Higher levels of group participation by an individual may translate into realization of a wide range of benefits, whereas lower levels of participation may signify a lack of effective access to benefits.
- *Trends in group participation.* Increase in group participation rates over time is consistent with a welfare-improving effect of groups. To assess whether there was growth in group memberships the respondent was asked: Compared to 3 years ago does your household today participate actively in *more*, *same* or *fewer* groups? The response was scored as 1 = more, 2 = same, and 3 = fewer. If more or less, the respondent was asked to indicate the reasons for the change.
- *Benefits accessed by group member.* This refers to the benefits participating individuals get by being members of a particular group. These include social services, economic opportunities (e.g. credit), extension services (e.g. agricultural technologies), empowerment (e.g. civic engagement) and natural resources management services. The benefits access variable was thus estimated by asking a surveyed household whether by being a member of a group it had obtained any of the benefits indicated in Table 9.1. A benefit accessed was coded 1, and otherwise zero. Based on an implicit assumption that the 12 diverse benefits are equally valuable to a household's well-being, a single value was obtained by summing all the benefits obtained from each type of group, with higher scores denoting a household enjoying greater access and diversity of benefits through the same type of group.[1]

**Table 9.1.** Benefits accessed through community and supra groups.

| | |
|---|---|
| 1. Education and/or training | 7. Agricultural technologies |
| 2. Health services | 8. Irrigation services |
| 3. Water and sanitation services (e.g. construction of wells) | 9. Land, forestry or water rights |
| | 10. Access to markets |
| 4. Electricity or solar energy | 11. Food and improved nutrition |
| 5. Credit/loans plans | 12. Civic education (e.g. political representation, conscious raising) |
| 6. Savings plans | |

[1] This implies a further assumption that benefits accessed through group participation are equally valuable in improving well-being.

### 3.3 Household welfare measures

Effective poverty reduction requires accurate measures of well-being (Barrientos and Hulme, 2005). By examining the relationship between membership in different group types – together with the benefits they leverage – and household welfare, we can estimate the welfare effects of household participation in these groups.

*Annual household income* and *ownership of assets* were used to estimate a household's economic well-being. Annual income based on estimates of the last 12 months prior to this study was computed from informal and formal income sources that included crop and livestock activities, non-farm activities and permanent employment of all family members living in the same household at the time of the interview.

*Ownership of assets* was determined by considering key household items and conditions of respondents' dwellings. The information was then used to compute an asset index using principal components analysis (PCA) (Stevens, 1996; Filmer and Pritchett, 2000).[2] The asset index is an alternative to income for understanding welfare differences, with a higher score denoting higher economic well-being. Sahn and Stifel (2003) posit two reasons for using asset-based measurements as appropriate measures of economic welfare. First, poverty alleviation is largely gauged against an individual's ability to acquire productive or income-generating assets, which are inherently a source of income inequality. Second, measurement problems such as reliance on recall data, inability to value the goods consumed from home production and poor-quality surveys make expenditure data unreliable.

Use of a household asset index in estimation of welfare allows for comparisons of wealth across households. We used principal component analysis on a set of household assets to construct the index. The procedure of PCA begins with a set of $K$ variables, $a^*_{1i}$ to $a^*_{Ki}$ representing the ownership of $K$ assets by the $i$th household. This is represented by binary scale: 1 if the asset $K$ is owned, and zero otherwise. Each variable $a^*_{1i}$ is specified by its mean and standard deviation, $a_{1i} = (a^*_{1i} - a^*_1)/(s^*_1)$, where $a^*_1$ is the mean of $a^*_{1i}$ across all $N$ households and $s^*_1$ is the standard deviation. The selected variables are linked with latent components (factors) for each household $i$ through the equation:

$$
\begin{aligned}
&a_{1i} = \upsilon_{11} \times A_{1i} + \upsilon_{12} \times A_{2i} + \cdots + \upsilon_{1K} \times A_{Ki} \\
&\ldots \qquad\qquad i = 1, \ldots, N \text{(households)} \\
&\qquad\qquad\quad k = 1, \ldots, K \text{(household assets)} \\
&a_{Ni} = \upsilon_{N1} \times A_{1i} + \cdots + \upsilon_{KK} \times A_{Ki} \qquad (9.1)
\end{aligned}
$$

where all $A$ are the components and all $\upsilon$ the coefficients on each component for each variable, and these are constant across all households. It is only the

[2] Filmer and Pritchett (2000) view asset index and expenditures as plausible proxies for something unobserved: a household's long-term economic status.

left-hand side that is observed, making the solution to the problem indeterminate. The PCA solves this by determining specific linear combinations of the variables with maximum variance accounted for in the first principal component $A_{1i}$ (Lawley and Maxwell, 1971). The procedure is repeated for each successive component accounting for the maximum of variance remaining. Reversing equation (9.1) yields factor loading from the model that are estimates for each of the $K$ principal components:

$$\begin{aligned} A_{1i} &= f_{11} \times a_{1i} + f_{12} \times a_{2i} + \cdots + f_{1K} \times a_{Ki} \\ &\ldots \qquad\qquad\qquad\qquad i = 1, \ldots, N \\ A_{Ki} &= f_{K1} \times a_{1i} + f_{K2} \times a_{2i} + \cdots + f_{KK} \times a_{Ni} \end{aligned} \tag{9.2}$$

where $A_{1i}$ is the first principal component, $a_{1i}$ the normalized variable, $f_{1i}$ is the factor score coefficient (weight) by which the normalized variable is multiplied to obtain a factor score in the linear combination. Thus, the asset index for each household is based on the expression:

$$A_{1i} = f_{11} \times (a^*_{1i} - a^*_1)/(S^*_1) + \cdots + f_{1N} \times (a^*_{ni} - a^*_N)/(S^*_N) \tag{9.3}$$

As many as 26 asset indicators were used in constructing the asset index[3]: (i) consumer durables – radio, television, bicycle, motor vehicle and mobile phone; (ii) farm investments, equipment and implements, particularly zero grazing unit, sewing machine, ox-cart, chaff cutter, knapsack sprayer, pruning equipment, wheelbarrow, folk hoes, spades; and (iii) quality of main dwelling's features: Roof, 0 = low and 1 = high; Walls, 0 = low and 1 = high, and Floor, 0 = low and 1 = high; Toilet facilities, 0 = low and 1 = high; Main source of cooking: Kenya ceramic stove, kerosene stove and electricity; Lighting: pressure lamp, solar energy and electricity; and Main source of water: 0 = spring or other sources and 1 = tap water.

Among the durables, ownership of television has the highest loading factor (0.633) while a bicycle has the lowest (0.318). Owning a zero grazing unit raises the index by 0.24 units while owning a forage cutter raises it by 0.45 units. Ownership of a vehicle or a mobile phone increases the asset index score by 0.31 and 0.24 units, respectively. The floor quality of the main house has the highest loading of 0.66 units and raises the asset index by 0.23 units. However, electricity increases the index most, by 0.34 units.

The asset index is divided into four equal-sized categories, or quintiles, with the poorest 25% of households designated quintile 1, and the richest 25%

[3] We started with 42 items but 16 of them did not show a high level of reliability as they had low-loading factors that were neither statistically nor significantly different from zero. Such items included tractor, landline phone, power saw, ox-plough, slashers, fireless cooker, lantern lamp, traditional charcoal stove, type of gate and type of fence. Those retained had a reliability of at least 0.8 and a loading of 0.3, a cut-off point meaningful for interpreting factors (Kline, 1994).

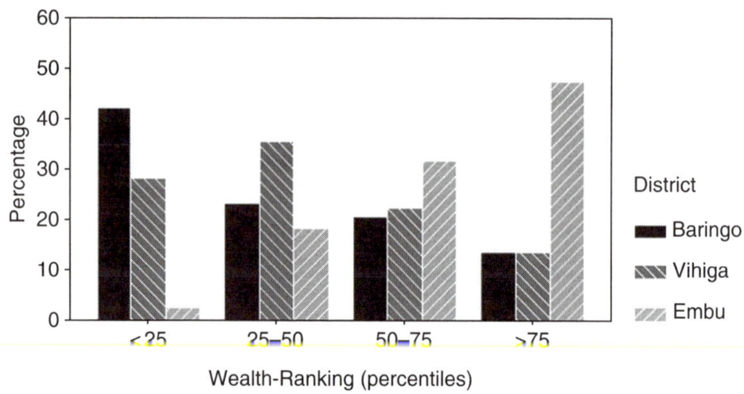


**Fig. 9.2.** Distribution of household wealth.

comprising quintile 4. The asset index differs statistically significantly across the three districts. Baringo has the highest proportion of its households, about 43%, ranking below the 25th percentile (Fig. 9.2). A number of Vihiga households (36%) fall in the 25–50th percentile, while a near-majority (47%) of the Embu households fall into the 50–75th percentile. Embu households appear generally wealthier than either Vihiga or Baringo households.

### 3.4 Descriptive statistics and unconditional analyses

Using *F*-tests for continuous variables and chi-square tests for categorical variables, the results show statistically significant differences in the means of the household variables across the three districts (Table 9.3). Most of the households (83%) are male-headed while the average age of the household heads is about 50 years.

A typical household consists of seven members with about 45% having at least someone engaged in off-farm employment. Average land size is 1.40 ha, with Baringo households having more than twice the size of Vihiga's and more than 25% Embu's. On average, 41% of Baringo household heads have a primary level of education while Vihiga has 33% and Embu 28%. However, a lower proportion in Baringo has attained secondary education, 36% compared to 53% and 54% in Vihiga and Embu, respectively.

A significant proportion of household heads in Baringo (22%) have completed tertiary education compared to 16% and 13% in Embu and Vihiga, respectively. On the other hand, Embu households have the highest average annual income, about KSh 125,600 compared with Baringo's KSh 71,000 and Vihiga's KSh 87,000. Results in Table 9.3 also show that in spite of Embu's relatively large proportion of respondents (34%) having access to financial institutions (ROSCAS and cooperative and village banks) compared with Vihiga's 30% and Baringo's 19%, more Vihiga respondents (61%) received

**Table 9.2.** Factor score coefficients and descriptive statistics for asset indices.

| Households owning assets (%) | Loading factor ($v_{1i}$) | Factor score coefficient ($f_{1i}$) | Mean | Standard deviation (SD) | Δ Asset (factor) index $f_{1i}$/SD |
|---|---|---|---|---|---|
| **Durables** | | | | | |
| Radio | 0.409 | 0.063 | 0.79 | 0.407 | 0.155 |
| Television | 0.633 | 0.097 | 0.22 | 0.414 | 0.234 |
| Bicycle | 0.318 | 0.049 | 0.40 | 0.490 | 0.100 |
| Vehicle | 0.370 | 0.057 | 0.04 | 0.185 | 0.308 |
| Mobile phone | 0.515 | 0.079 | 0.13 | 0.336 | 0.235 |
| **Farm investments and equipment** | | | | | |
| Zero grazing unit | 0.472 | 0.073 | 0.11 | 0.311 | 0.235 |
| Chaff/forage cutter | 0.327 | 0.050 | 0.01 | 0.111 | 0.450 |
| Ox-cart | 0.425 | 0.065 | 0.08 | 0.270 | 0.240 |
| Knapsack-sprayer | 0.658 | 0.101 | 0.22 | 0.417 | 0.242 |
| Wheelbarrow | 0.702 | 0.108 | 0.28 | 0.448 | 0.241 |
| Fork hoe | 0.540 | 0.083 | 0.55 | 0.498 | 0.167 |
| Spades | 0.467 | 0.072 | 0.59 | 0.49 | 0.147 |
| Sewing machine | 0.384 | 0.059 | 0.10 | 0.303 | 0.195 |
| Hand saw | 0.521 | 0.080 | 0.29 | 0.454 | 0.176 |
| Pruning equipment | 0.600 | 0.092 | 0.27 | 0.445 | 0.207 |
| **Characteristics of dwellings** | | | | | |
| Kenya ceramic stove | 0.386 | 0.059 | 0.23 | 0.418 | 0.141 |
| Electric/gas cooker | 0.433 | 0.067 | 0.04 | 0.200 | 0.335 |
| Kerosene stove | 0.487 | 0.075 | 0.35 | 0.479 | 0.157 |
| Cereal store | 0.303 | 0.047 | 0.27 | 0.446 | 0.105 |
| Pressure lamp | 0.466 | 0.072 | 0.10 | 0.300 | 0.240 |
| Tap water | 0.532 | 0.082 | 0.24 | 0.429 | 0.191 |
| Solar energy | 0.524 | 0.081 | 0.11 | 0.316 | 0.256 |
| Roof of main house | 0.415 | 0.064 | 0.82 | 0.386 | 0.166 |
| Walls of main house | 0.479 | 0.074 | 0.17 | 0.377 | 0.167 |
| Floor of main house | 0.660 | 0.102 | 0.27 | 0.446 | 0.229 |
| Type of toilet | 0.652 | 0.100 | 0.13 | 0.336 | 0.298 |
| **Asset index** | | | 0.00 | 1.00 | |

Notes: The variance explained by the assets on the first principal component is 25.0% with eigenvalue 6.5.

credit compared to Embu's (49%) and Baringo's (43%). This suggests that knowledge of credit as a source of funds for investment, rather than consumption, has lately grown in Vihiga. As would be expected, Baringo, which largely relies on a livestock economy, has almost three times and four times the TLUs[4] as Vihiga's and Embu's, respectively.

For the main cash crops, about 52% of Embu farmers grow tea, twice as many as in Vihiga. Embu district also leads in coffee with 69% of the respond-

[4] TLU is tropical livestock unit (1 TLU = 250 kg), with camel = 1, bull = 0.7, cow = 0.6, steer = 0.5, heifer = 0.4, calf = 0.2, shoat = 0.1, lamb or kid = 0.03 (ILCA, 1990).

**Table 9.3.** Descriptive characteristics of sample households ($N = 480$).

| Variable | Baringo | Vihiga | Embu | Overall | Sig. |
|---|---|---|---|---|---|
| 1. Gender | | | | | |
| (a) Male-headed | 83.8 | 83.1 | 83.1 | 83.3 | *** |
| (b) Female-headed | 16.3 | 16.9 | 16.9 | 16.7 | |
| 2. Age of household head (years) | 44.8 | 51.5 | 52.1 | 49.5 | *** |
| 3. Highest education of hh (%) | | | | | |
| (a) Non-formal | 1.3 | 0.7 | 1.9 | 1.04 | ** |
| (b) Primary | 40.6 | 33.1 | 28.1 | 34.0 | |
| (c) Secondary | 35.6 | 53.1 | 53.8 | 47.5 | |
| (d) Tertiary | 21.3 | 13.1 | 16.2 | 16.6 | |
| 4. Off-farm work | | | | | |
| (a) % households without | 75.5 | 43.8 | 47.5 | 55.5 | *** |
| (b) % households with | 24.5 | 56.2 | 52.7 | 44.5 | |
| 5. Household size (persons) | 7.0 | 7.8 | 6.6 | 7.1 | *** |
| 6. Land size (ha) | 1.95 | 0.75 | 1.50 | 1.40 | *** |
| 7. Title deed (% possesses) | 33.7 | 39.9 | 63.5 | 45.7 | *** |
| 8. Total livestock unit (TLU) | 3.72 | 1.39 | 0.98 | 2.03 | *** |
| 9. Gross income (KSh, '000) | 70.9 | 86.7 | 125.6 | 52.0 | ** |
| 10. Obtained credit (%) | 43.1 | 61.3 | 48.8 | 54.3 | |
| (a) ROSCAs | 3.8 | 16.9 | 2.5 | 7.7 | *** |
| (b) Cooperative | 14.4 | 10.6 | 20.0 | 15.0 | * |
| (c) Village bank | 1.2 | 2.5 | 11.3 | 5.0 | *** |
| 11. Crops (% households growing) | | | | | |
| (a) Tea | 0.0 | 25.0 | 51.9 | 25.5 | *** |
| (b) Coffee | 16.3 | 11.3 | 68.8 | 32.1 | *** |
| (c) Maize | 94.3 | 100 | 98.1 | 97.5 | ** |

Exchange rate: 1 KSh = US$75.
Statistically significant differences across districts at $^{*}p < 0.05$; $^{**}p < 0.01$; $^{***}p < 0.001$.

ents growing the crop, compared with Vihiga's 11% and Baringo's 16%. Maize is a staple crop grown by almost every household in the three districts.

## 3.5 Group participation patterns

Table 9.4 shows that there is considerable differentiation in group participation across districts.[5] About 29%, 21% and 19% of households do not participate in any local group in Baringo, Vihiga and Embu, respectively. However, about 71%, 77% and 74% of the households participate in 1–3 groups in Baringo,

[5] Effect size (phi) is a statistic in the form of correlation in standard units that reflects the amount of variation in the dependent variable that is linked to the independent variable, in this case group membership and district, respectively. As a guide, effect size is interpreted as: <0.1 weak, <0.3 modest, <0.5 moderate, <0.8 strong and ≥0.8 very strong. Values below 0.1 and above 0.9 indicate that the independent variable accounts for very little and almost all variance respectively, in the dependent variable (De Vaus, 1991; Kline, 1994).

**Table 9.4.** Household memberships in community and supra groups (%).

| Group type | Group membership | Baringo | Vihiga | Embu | Total |
|---|---|---|---|---|---|
| Community groups | 0 | 29.4 | 20.6 | 18.8 | 22.9 |
| (chi-square = 73.5, | 1 | 58.8 | 48.1 | 27.5 | 44.8 |
| df = 12, $p < 0.001$, | 2 | 10.0 | 21.9 | 30.0 | 20.6 |
| phi = 0.391) | 3 | 1.9 | 6.9 | 17.5 | 8.8 |
| | ≥4 | 0.0 | 2.5 | 6.4 | 2.9 |
| Supra groups (chi- | 0 | 60.0 | 8.8 | 9.4 | 26.0 |
| square = 283.78, | 1 | 30.6 | 58.8 | 12.5 | 34.0 |
| df = 14, $p < 0.001$, | 2 | 6.9 | 21.9 | 26.3 | 18.3 |
| phi = 0.769) | 3 | 2.5 | 8.1 | 41.3 | 17.3 |
| | ≥4 | 0.0 | 2.5 | 10.7 | 4.3 |

Vihiga and Embu, respectively. A similar pattern is observed in supra groups, although these are far less common in Baringo than in the highlands sites. As many as 60%, 9% and 9% do not participate in any supra group in Baringo, Vihiga and Embu, respectively, while 40%, 89% and 80% participate in 1–3 supra groups, respectively. While 11% of the Embu households participate in at least four supra groups, only 3% of Vihiga's do.

The results also suggest that most households belong to a group but very few, except Embu farmers, belong to more than three supra groups. In general, the level of group participation is highest in Embu compared with Vihiga and Baringo. This suggests that households in Embu experience far greater benefits from group participation, consequently acting as a motivation for joining or forming groups. The findings suggest that Embu district is endowed with factors conducive to the development of supra groups such as proximity to the capital or political influence, and infrastructure. This suggests the need for initiating development programmes in helping foster middle-level institutions, including village elders, with potential to help rural people in remote locations who suffer from chronic poverty, chronic illnesses or historical marginalization break free of the economic and cultural constraints on their struggles out of poverty.

### 3.6 Trends in group participation

There are significant differences in group membership growth with modest effect (phi = 0.319) across the three districts over the last 3 years (Fig. 9.3). Examining group participation over long-term periods helps unravel the influence of social dynamic processes that sustain or curtail social capital necessary for allowing rural communities reap economic advantages – in terms of livelihood stability, employment generation, poverty reductions, quality of services, among others. Such economic functions are contingent upon high intra-village social capital on one hand, and ready access to agents capable of targeting this capital towards incentives arising from the external environment on the other

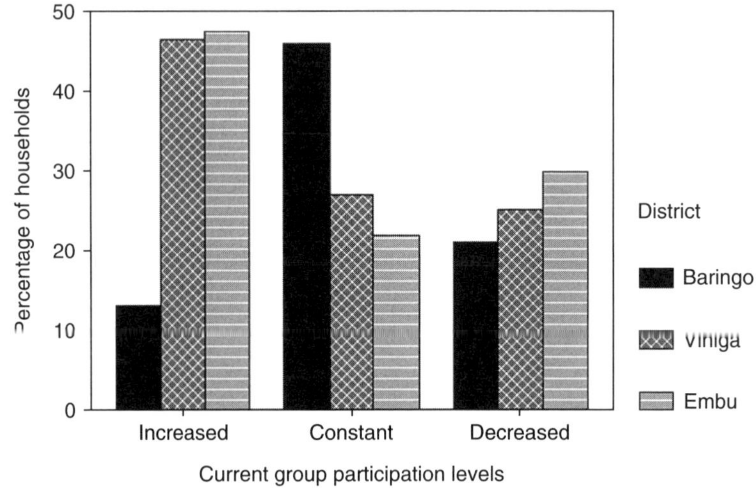


**Fig. 9.3.** Trends in household group participation in 2000–2003.

(Krishna, 2003). Figure 9.3 shows that participation in groups in Embu and Vihiga has increased significantly while participation in Baringo has increased marginally over the last 3 years. A total of 48% of the Embu households report participating in more groups, compared with 47% of Vihiga and 21% of Baringo respondents. The converse is observed in group membership stagnation, with 22% of households in Embu, 27% in Vihiga and 46% in Baringo indicating no growth. Less than 5% of the respondents did not respond to this item.

These findings suggest that groups in Embu and Vihiga continue to be relevant in providing households with access to key welfare-enhancing resources. However, increased group participation in Embu seems to be associated with more effective social capital than in Vihiga and Baringo. For example, results in Table 9.5 show significant differences in level of social capital that can be drawn on in times of need. More households in Embu (50%) than in Vihiga (29%) or Baringo (21%) indicated assurance of social support from a group whenever

**Table 9.5.** Social context for group participation (% of respondents).

| Social capital dimension | | Baringo | Vihiga | Embu | Overall |
|---|---|---|---|---|---|
| Is there an individual you can rely on for help? (chi-square = 72.03, df = 6, $p < 0.01$, phi = 0.388) | NA | 1.3 | 0.0 | 0.0 | 0.4 |
| | Definitely 1 | 54.7 | 39.4 | 62.9 | 52.3 |
| | Probably 2 | 37.7 | 20.0 | 20.8 | 26.2 |
| | Probably not 3 | 6.3 | 40.6 | 16.4 | 21.1 |
| Is there a group you can rely on for help? (chi-square = 101.8, df = 6, $p < 0.01$, phi = 0.461) | NA | 0.6 | 0.0 | 0.0 | 0.2 |
| | Definitely 1 | 21.3 | 29.4 | 50.3 | 33.6 |
| | Probably 2 | 59.4 | 18.1 | 23.9 | 33.8 |
| | Probably not 3 | 18.8 | 52.5 | 25.8 | 32.4 |

a household is very short of money and food, or if members of the household were ill. Similarly, about 63% of the Embu participants compared with about 55% of Baringo and 39% of Vihiga participants were definitely assured of support from an individual whenever a member sought for assistance outside the household or other than relatives.

The results suggest that in spite of the higher group memberships by households in Vihiga, the social capital manifested responds less to needy cases than in Embu and Baringo. This suggests that group participation may not necessarily foster effective social security networks for mutual insurance in some areas. The results may also be interpreted to imply that different groups may encourage members to exercise their individual and group capabilities in different ways. For example, some groups may place more value on individualism than on social interaction, leading to what Putnam (2000) calls the 'bowling alone' syndrome.

Putnam *et al.* (1993) and Putnam (1996) argue that poor neighbourhoods that lack the necessary qualities of self-help, mutuality and trust tend to generate limited benefits, hence providing less support in critical times. Similarly, a poorly endowed group may be limited in its capacity to generate benefits, and thus the provision of additional resources may improve the quality of social capital for such a group. Neighbourhoods in which existing relations of trust and reciprocity are weak tend to lack the qualities that create and sustain voluntary associations and partnerships (Trigilia, 2001). On the other hand, communities in which there is strong attachment to groups also generate the energy to contribute to the group's activities, which lead to anticipated benefits. These include contribution of money, labour or materials (Prokopy, 2005), and involvement in higher levels of decision making (Schouten and Moriarty, 2003).

## 3.7 High participation in community groups

Table 9.6 presents 13 responses provided to an open-ended question by the respondents who indicated that they had increased participation in more groups compared to 3 years ago. A careful examination of the factors[6] extracted by the PCA suggests that individuals join groups due to both self- and collective interests.

A majority of the variance (19.2%) is explained by factor 1, which, given the factor loadings, can be associated with a public good incentive for group participation, which includes collective action, learning from others, problem solving and maintaining a social order through norm enforcement. On the other hand, factor 2 (12.5%) is associated with a private incentive for joining groups such as improving the standard of living, engaging in diversified income-generating

[6] A rotation (e.g. Varimax) is made to obtain factors which load high on only a small number of variables with the rest being close to zero (Stevens, 1996). This makes the use of maximum variance property not reliably meaningful, but facilitates interpretation once the amount of variance accounted for by each rotated factor is recalculated.

**Table 9.6.** Reasons for high level of participation in groups.

| | Factor pattern | | | |
|---|---|---|---|---|
| Variable | 1 | 2 | 3 | 4 |
| Collective action in public goods | **0.709** | 0.150 | –0.107 | –0.073 |
| Education enlightenment | **0.681** | 0.252 | 0.004 | –0.057 |
| Improves supply of social capital | **0.677** | 0.280 | 0.276 | 0.126 |
| Problem solving and social support | **0.676** | 0.128 | 0.386 | –0.077 |
| Enhance social norms for social control | **0.563** | –0.240 | 0.065 | 0.200 |
| Generate cash flows – savings, loans, etc. | 0.279 | 0.610 | 0.005 | 0.117 |
| Improve standards of living | 0.366 | **0.607** | –0.133 | 0.304 |
| Enhance efficiency on farm | 0.031 | **0.533** | 0.220 | –0.001 |
| Enhance income-generating activities | –0.025 | **0.477** | 0.348 | –0.221 |
| Enhance welfare development | 0.285 | –0.025 | **0.795** | –0.021 |
| Acquire productive assets | –0.055 | 0.327 | **0.648** | 0.127 |
| Gain access to markets | 0.020 | –0.045 | 0.234 | **0.795** |
| Reduce poverty | –0.008 | 0.155 | –0.160 | **0.760** |
| Eigenvalue (7.14) | 3.31 | 1.45 | 1.32 | 1.06 |
| % variance explained (54.9) | 19.2 | 12.5 | 12.0 | 11.2 |

Note: Extraction method: principal component analysis. Rotation method: Varimax with Kaiser normalization.

activities and enhancing efficiency. Income-generating activities included poultry keeping, fruit processing, tea growing, production of high-value horticultural crops, dairy keeping and bee-keeping. Although factor 3 underscores welfare development and acquisition of productive assets and factor 4 underlies issues of markets and poverty reduction, both are proxies of welfare improvement.

These extracted factors suggest that the underlying incentives for group participation are related to collective, instrumental and practical benefits. Consequently, the propensity to participate in more groups is not only a source of improved capabilities, but a capability itself. This is especially so where relationships and networks sustain the expectations, norms and trust that facilitate the attainment of benefits. In spite of expected benefits from increased group participation, there has also been declining memberships in groups by some households.

## 3.8 Low participation in community groups

One can take a different approach to this question by exploring instead the correlates of low participation in community groups. PCA yielded a five-factor loadings structure of reasons underlying the decreased group participation (Table 9.7). Factor 1 (17%) is associated with poor management, lack of social cohesion, lack of accountability, poor organizational arrangements and logistics, and lack of support to group leadership by some members. Factor 2 (12.4%) is associated with lack of financial resources and high membership demands; this factor is undoubtedly linked to differences in income. Factor 3 (10.7%) indicates dependency on exter-

**Table 9.7.** Constraints to participation in groups.

| Variable | Factor 1 | 2 | 3 | 4 | 5 |
|---|---|---|---|---|---|
| Poor management | **0.733** | 0.057 | 0.191 | 0.025 | −0.043 |
| Misunderstandings (lack of transparency) | **0.745** | 0.013 | −0.027 | 0.248 | 0.157 |
| Lack of accountability | **0.715** | 0.073 | −0.020 | 0.163 | 0.076 |
| Poor arrangement and logistics | **0.503** | 0.163 | 0.095 | −0.330 | −0.470 |
| Lack of support for leadership | **0.528** | 0.329 | 0.084 | 0.157 | 0.137 |
| Lack of financial resources | 0.274 | **0.702** | 0.289 | 0.027 | −0.029 |
| High share contributions | 0.036 | **0.752** | −0.131 | 0.007 | 0.382 |
| Lack of active membership | 0.076 | **0.518** | −0.064 | −0.084 | −0.278 |
| Dependency on external help | 0.115 | 0.089 | **0.573** | −0.202 | 0.274 |
| Lack of commitment | −0.121 | 0.484 | **0.556** | 0.347 | −0.161 |
| Heterogeneity of members | 0.091 | −0.083 | **0.804** | 0.046 | 0.022 |
| Groups not helpful – seek alternative income generation | 0.194 | 0.096 | 0.067 | **0.728** | −0.173 |
| Lack of clear goals and objectives | 0.219 | −0.090 | −0.089 | **0.672** | 0.180 |
| Lack of self-expression | 0.298 | 0.029 | 0.230 | −0.073 | **0.650** |
| Eigenvalue (8.14) | 3.11 | 1.57 | 1.25 | 1.20 | 1.01 |
| % variance explained (57.6) | 16.9 | 12.4 | 10.7 | 9.9 | 7.7 |

Extraction method: principal component analysis. Rotation method: Varimax with Kaiser normalization.

nal support, lack of commitment due to domestic chores and heterogeneity among group members. Factor 4 (10%) underlies lack of benefits from groups warranting alternative income-generating activities, and a lack of clear goals and objectives. As factor 5 adds relatively little to explaining the common variance, it is set aside.

Given that factor 1, which explains the majority of the variance in reasons cited for low group participation, is associated with leadership and management functions, the results point to the need for leadership that is transparent, accountable and one that has effective organizational and managerial capacity to keep every member actively involved in the network. Additionally, processes and activities that could boost groups' financial contributions may enhance committed memberships. Thus, the groups may self-select those individuals who are more proactive, energetic and resourceful, leaving behind those who are much poorer and perhaps at greater risk and need. Importantly, individuals who aim to move up an economic pyramid structure that is characterized by different levels of financial requirements may be unable to realize their goals, yet such layers of commitment are necessary for active memberships.

## 3.9 Benefits accessed through community and supra groups

A one-way analysis of variance of means of indices representing benefits households derive from both community and supra groups shows statistically significant differences from zero across the three districts (Table 9.8).

**Table 9.8.** Comparisons of Means of Benefits Access Index (MBAI).

| Dependent variable | (I) District | MBAI | (J) District | MBAI | Mean difference (I–J) | Std Error |
|---|---|---|---|---|---|---|
| Benefits accessed by community groups | Vihiga | 3.73 | Baringo | 1.43 | 2.30(*) | 0.55 |
| | Embu | 6.08 | Baringo | 1.42 | 4.66(*) | 0.55 |
| | Embu | 6.08 | Vihiga | 3.72 | 2.36(*) | 0.55 |
| Benefits accessed by supra groups | Vihiga | 3.36 | Baringo | 1.26 | 2.09(*) | 0.51 |
| | Embu | 9.18 | Baringo | 1.26 | 7.92(*) | 0.51 |
| | Embu | 9.18 | Vihiga | 3.36 | 5.82(*) | 0.51 |

*The mean difference is significant at the 0.05 level.
Benefits accessed by community groups across districts: $F = 35.802$, df = 2, $p < 0.001$.
Benefits accessed by supra groups across districts: $F = 127$, df = 2, $p < 0.001$.

This suggests significant differences among districts in the benefits enjoyed from both types of groups. Embu households enjoyed about four times as many benefits through community groups as Baringo households and 63% more than Vihiga's. These differences were magnified in the case of supra groups, with Embu enjoying more than seven and almost three times as many benefits through supra groups as Baringo and Vihiga households, respectively. The male-headed households enjoyed 50% more benefits from supra groups than did female-headed households (means: M-headed = 5.4, F-headed = 3.6, $F = 3.00$, $p < 0.1$). Moreover, the benefits of group participation were enjoyed more by the rich than the poor. Table 9.9 shows that the rich (those in the sample above the 50th percentile of the wealth ranking) benefited from supra group participation by more than 61% that of the poor (those below the 50th percentile). The effects are magnified further in the benefits enjoyed through the same groups, with the wealthier getting services more than 100% those by the poor. These results suggest that supra group participation generates spatially uneven effects and that the resulting inequality could be linked to various structural inequalities, e.g. wealth, ethnicity (the three districts have three different main ethnic communities).

**Table 9.9.** Comparisons of supra group participation across wealth categories.

| | Poor | Rich | *F*-test |
|---|---|---|---|
| Independent variable | Mean Asset Index below 50th percentile | Mean Asset Index above 50th percentile | |
| Average number of supra groups | 1.09 | 1.76 | $F = 37.45$, df = 2 $p < 0.01$ |
| Average benefits accessed through supra groups | 2.98 | 6.28 | $F = 37.45$, df = 2 $p < 0.01$ |

# 4 The Effects and Determinants of Group Participation

Ultimately, this research seeks to establish the impact of group participation, differentiated between community groups and supra groups, and the benefits accessed through them on household-level well-being. Since group membership is non-random, we also seek to understand how patterns of group participation vary within rural communities in Kenya, especially in so far as different sorts of groups may generate different levels or types of benefits to members.

## 4.1 The differentiated impact of groups on household well-being

We start by exploring the impact of group membership on our two measures of household well-being: assets and income. An ordinary least squares (OLS) multiple regression was estimated to explain the relationship between household welfare and group participation measures:

$$Y = b_0 + b_1x_1 + b_2x_2 + b_3x_3 + \cdots + b_{12}x_{12} + \varepsilon \quad (9.4)$$

where $Y$ is a measure of household-level well-being, either the asset index defined earlier or the natural logarithm of estimated annual household income. The independent variables are *gender* of household head (hh) (1 = male, 0 = female); *age* of hh in years; *highest education level* attained by the hh (1 = non-formal, 2 = primary, 3 = secondary, 4 = college and tertiary education); *size of household* measured by the number of household members living in the home; *remittance potential* measured by the number of household members employed off the farm; *land size* in hectares; *land tenure* measured by ownership of title deed (yes = 1, otherwise 0); *total livestock units* (TLUs) measured as an index constructed from numbers and types of livestock[7]; *group memberships* measured by the number of either local or supra groups that a household is a member of; *benefits access* measured as defined earlier; and *district type* from which a surveyed household hails (two dummies: Vihiga = 1, otherwise 0; and Embu = 1, otherwise 0; Baringo is the reference district).

Do the density of group memberships and the quantity of benefits accessed through them measurably affect household welfare? Using the OLS regression analyses to test this hypothesis, the independent variables account for 54% of the variance in the asset index and 32% of the variance in the log of income (Table 9.10).

The results show that household resource endowments – level of education, size of livestock, size of land and secure land tenure – have the expected, significant and positive effects on the two welfare measures. Similar effects are also associated with young, male-headed households and residents in either

[7] This variable indicates the size of livestock holding that a household owns and is derived from various conversions: 1 Tropical Livestock Unit = 250 kg, with bull = 0.7 LU, cow = 0.6 LU, steer = 0.5 LU, heifer = 0.4 LU, calf = 0.2 LU, shoat = 0.1 LU, lamb or kid = 0.03 LU. Tropical livestock unit is typically taken to be equivalent to 250 kg of annual weight and is unitless (ILCA, 1990).

**Table 9.10.** Results of multiple regression analyses for welfare measures ($N = 480$).

| Dependent variables | Asset index log income | | | | | |
|---|---|---|---|---|---|---|
| Independent variables | B | SD | Beta | B | SD | Beta |
| (Constant) | −1.156 | 0.152 | *** | 9.635 | 0.239 | *** |
| Gender of household head (1 = Male, 0 = Female) | 0.347 | 0.088 | 0.129*** | 0.363 | 0.139 | 0.106*** |
| Age of household head (years) | −0.006 | 0.002 | −0.091** | −0.017 | 0.004 | −0.196*** |
| Secondary education (1 = Y, 0 = N) | 0.267 | 0.072 | 0.133*** | 0.348 | 0.114 | 0.136*** |
| Tertiary education (1 = Y, 0 = N) | 0.781 | 0.103 | 0.292*** | 0.938 | 0.163 | 0.273*** |
| Size of household | −0.016 | 0.013 | −0.047 | 0.015 | 0.020 | 0.034 |
| Household members in off-farm employment | 0.056 | 0.028 | 0.074** | 0.139 | 0.043 | 0.145*** |
| Land size (ha) | 0.108 | 0.020 | 0.184*** | 0.137 | 0.032 | 0.183*** |
| Title deed (1 = Y, 0 = N) | 0.413 | 0.074 | 0.205*** | 0.140 | 0.116 | 0.054 |
| Total livestock units (TLU) | 0.041 | 0.010 | 0.148*** | 0.052 | 0.015 | 0.151*** |
| Memberships in local groups | 0.046 | 0.038 | 0.049 | 0.060 | 0.060 | 0.050 |
| Memberships in supra groups | 0.124 | 0.041 | 0.150*** | 0.203 | 0.065 | 0.193*** |
| Benefits from local groups | 0.013 | 0.008 | 0.066 | 0.002 | 0.013 | 0.008 |
| Benefits from supra groups | 0.022 | 0.009 | 0.125** | 0.012 | 0.014 | 0.052 |
| Vihiga (1 = Y, 0 = N) | 0.031 | 0.090 | 0.014 | 0.404 | 0.143 | 0.149*** |
| Embu (1 = Y, 0 = N) | 0.491 | 0.111 | 0.231*** | 0.423 | 0.176 | 0.156** |
| Adjusted $R^2$ | | | 0.542 | 0.318 | | |
| Standard error of welfare measure | | | 1.0 | 1.28 | | |

Note: [a]$p < 0.10$; **$p < 0.05$; ***$p < 0.01$.
B and Beta are the unstandardized and standardized regression coefficients, respectively.

Embu or Vihiga. However, the age of the household head has a negative and statistically significant effect on both welfare measures. Size of household is not statistically significant, but households with higher potential for supplementary income, as reflected by a larger size of members in off-farm employment, are more likely to be well off.

Land is a key resource for rural households. Having more land can enhance farm production, consequently increasing the likelihood of acquiring other assets of much higher value and raising household incomes. De Janvry *et al.* (2001) suggest that access to land with secure tenure could be used as: (i) store of wealth against inflation; (ii) source of self-employment and food security if there are labour market and food market failures; (iii) collateral for credit; (iv) source of insurance to access credit subsidies; and (v) base for political and social capital. Similarly, livestock holdings are an important resource for rural households. The contrast between community groups and supra groups in the estimation is noteworthy. Controlling for the household and district variables, the number of supra groups significantly and positively affects both measures of household welfare. In addition, the corresponding benefits accessed through them only have significant effects with the asset index. Benefits accessed through supra groups are indicative of groups that meet the needs of households and improve their quality of life. Generally, supra groups are linked to higher-level organizations leading to activities and processes that generate outcomes to members.

Perhaps surprisingly, given the widespread celebration of local groups (Bebbington, 1999; Narayan and Pritchett, 2000; Weinberger and Jüting, 2000; Lyon, 2003), there is at best only a weak link between the level of participation in local groups and income and asset measures of welfare. Although factor analyses suggest that local groups may enhance social capital and generate benefits, the benefits are not sufficient to significantly change welfare consistently and substantively as supra groups do. Local groups are therefore limited in expanding the realm of human agency necessary to influence the social, political and economic freedom of people. On the other hand, qualitative evidence from the study's focus group discussions suggests that local groups such as the women groups and self-help groups are viewed by members as effective entry point for the poor to access resources, but this is not sufficient.

These findings also suggest that better livelihood will have to come from expansion of supra group involvement and substantial improvements in their benefits generation. This requires effective linkages with external organizations and improving access to more productive assets and activities. Some researchers indicate that such efforts need some combination of: (i) improved access for the poor to an array of productive assets; (ii) increased returns from the assets that they hold to strengthen livelihood security; and (iii) reduced vulnerability to natural and policy shocks (Ellis, 2000; Barrett *et al.*, 2001). The results suggest that for groups to leverage diverse benefits to members and the community in general, they require different forms of capital such as human and financial. A major source of capital for rural communities is credit from financial lending institutions (La Ferrara, 2002; Lyon, 2003). Hence, an effective group strategy that focuses on relevant key resources for the rural poor should seek to support groups where they are not efficient, expand their productive and useful functions, and promote access to credit and extension services. Such support could enable rural communities to learn how resources, information, decision making, delivery mechanisms and accountability can be structured to improve benefits (Pritchett and Woolcock, 2002). Moreover, access to information and knowledge from outside group boundaries may also enhance adaptation to turbulent and uncertain environments (Conway, 1995).

## 4.2 Determinants of participation in community and supra groups

We also seek to understand patterns of group participation. We approach this question by estimating the logistic regression function:

$$Y_i = b_0 + b_{i1}x_{i1} + b_{i2}x_{i2} + \cdots + b_{i11}x_{i11} + e_i \qquad (9.5)$$

where $Y_i = 1$ if a household belongs to a certain group type (and equals zero otherwise). If different groups generate different levels of welfare benefits, we want to know whether all households are equally likely to belong to each type of group or, as is more likely, which types of households are most likely to belong to the groups that confer the greatest benefits.

So, do those who participate in supra and community groups have more assets or incomes than those who do not? The multiple logistic regressions of group participation on a suite of explanatory variables correctly predict more than 78% of the observed variation in comparing non-participation and participation (community groups: 78.1%; supra groups: 85.5%) (Table 9.11).[8]

The results indicate that neither assets nor income are statistically significantly associated with household participation in community groups. However, both measures of welfare are positively and significantly associated with the likelihood of a household participating in supra groups. Income increases the

**Table 9.11.** Fitted logistic model to local group and supra group participation.

| | Local group participation | | | Supra group participation | | |
|---|---|---|---|---|---|---|
| Independent variable | B | SE | Exp(B) | B | SE | Exp(B) |
| Gender of household head (1 = M, 0 = F) | −0.558* | 0.296 | 0.572 | −0.622* | 0.370 | 0.537 |
| Age of household head (years) | 0.000 | 0.009 | 1.000 | −0.002 | 0.011 | 0.998 |
| Secondary education (1 = Y, 0 = N) | −0.135 | 0.266 | 0.874 | −0.202 | 0.312 | 0.817 |
| Tertiary education (1 = Y, 0 = N) | −0.369 | 0.419 | 0.691 | −0.024 | 0.476 | 0.976 |
| Size of household | 0.058 | 0.047 | 1.059 | 0.086 | 0.058 | 1.090 |
| Household members in off-farm employment | −0.087 | 0.101 | 0.917 | 0.044 | 0.155 | 1.045 |
| Land size (ha) | −0.087 | 0.068 | 0.916 | −0.223*** | 0.075 | 0.800 |
| Has title deed (1 = Y, 0 = N) | 0.334 | 0.282 | 1.397 | −0.358 | 0.348 | 0.699 |
| Total livestock units | 0.045 | 0.045 | 1.046 | −0.037 | 0.033 | 0.964 |
| LOGINCOME | −0.016 | 0.112 | 0.984 | 0.237* | 0.125 | 1.267 |
| ASSET INDEX | 0.148 | 0.184 | 1.160 | 0.716*** | 0.233 | 2.047 |
| VIHIGA(1) | −0.557* | 0.317 | 0.573 | −2.490*** | 0.386 | 0.083 |
| EMBU(1) | −0.730** | 0.356 | 0.482 | −1.922*** | 0.401 | 0.146 |
| Constant | 2.332 | 1.594 | 10.301 | 2.281 | 1.787 | 9.783 |
| -2loglikelihood | 469.672 | | | 344.139 | | |
| Hosmer–Lemeshow | | | | | | |
| Test: (a) chi-square | 6.195 | | | 9.717 | | |
| (b) *p* value | 0.625 | | | 0.285 | | |
| Percentage correct prediction | 78.1 | | | 85.5 | | |

Statistically significant levels: *$p < 0.10$; **$p < 0.05$; ***$p < 0.01$.

[8] The low chi-square statistic values and the large Hosmer–Lemeshow $p$ values of $p > 0.150$ (supra = 0.285 and local = 0.625) indicate that the two models are adequate for linking the covariates with the conditional mean of the logit function.

odds ratio of participation in supra groups by a factor of about 3.6 ($e^{1.267}$) times per unit increase in natural logarithm of income and the ownership of assets increases the same type of participation by 7.7 ($e^{2.047}$) times per unit increase in the asset index. This can be interpreted as meaning that moving from the lowest-income quintile (0–20%) to the middle quintile (40–60%) increases the odds ratio of participation in supra groups by 1.4 times, and the same transition in assets increases the odds ratio of the same type of participation by 3.1 times. This suggests that increased assets have a much bigger effect on enhancing opportunities for participation in supra groups. Land, an indicator of wealth, however, has a negative effect on the odds of participation in supra groups.

The results from the multivariate linear regression and the ordinary logistic regression suggest that causality in the supra group participation and household welfare could be bidirectional. On the one hand, participation in supra groups may generate outcomes which positively influence the well-being of a household. On the other hand, as the potential of supra groups for benefits get recognized, those with greater resources (assets and incomes) are better able to incur upfront costs required in lieu of the anticipated benefits, and are hence more likely to increase their memberships in supra groups. Such groups support productive purposes such as livestock improvement, e.g. Dairy Goats Association of Kenya (DGAK) and cash crop farming, e.g. Kenya Tea Development Authority (KTDA) for tea and Coffee Saccos for coffee. The latter interpretation suggests that the poor face resource constraints to participation compared with the rich who own a range of assets and are able to draw on their economic resources to invest in supra groups. This is consistent with the large and statistically significant coefficient of asset index ($b = 0.716$) on the likelihood of joining supra groups compared with the coefficient of income ($b = 0.237$). These findings also corroborate the factor analysis results for low group participation where factors 2 and 4 are significantly associated with upfront costs such as subscription fees, upfront cash contributions and high share savings.

Although community groups have been regarded as possible vehicles for poverty reduction in developing rural areas, this study cautions against generalizations. How then do we explain these results? Qualitative explorations suggest that some community groups contribute to higher levels of household welfare presumably because of linkages with higher-level organizations, which enable households to access resources they otherwise would not. One major factor cited as impeding this level of participation is lack of financial resources to meet membership conditions. To optimally benefit from groups, interested members commit their resources, meaning that those who are able to do so are most likely to enjoy the value of productive groups. Figure 9.4 illustrates this point drawn from the savings and credit data of a poultry women's group in Embu district. As an example of a community group which is linked to some supra organizations such as the Kenya Rural Enterprise Programme (KREP) and the Kenya Women Finance Trust (KWFT), members who have the ability to pay entry and regular share contributions join the category that contributes higher shares to raise sufficient funds used as collateral for borrowing bigger

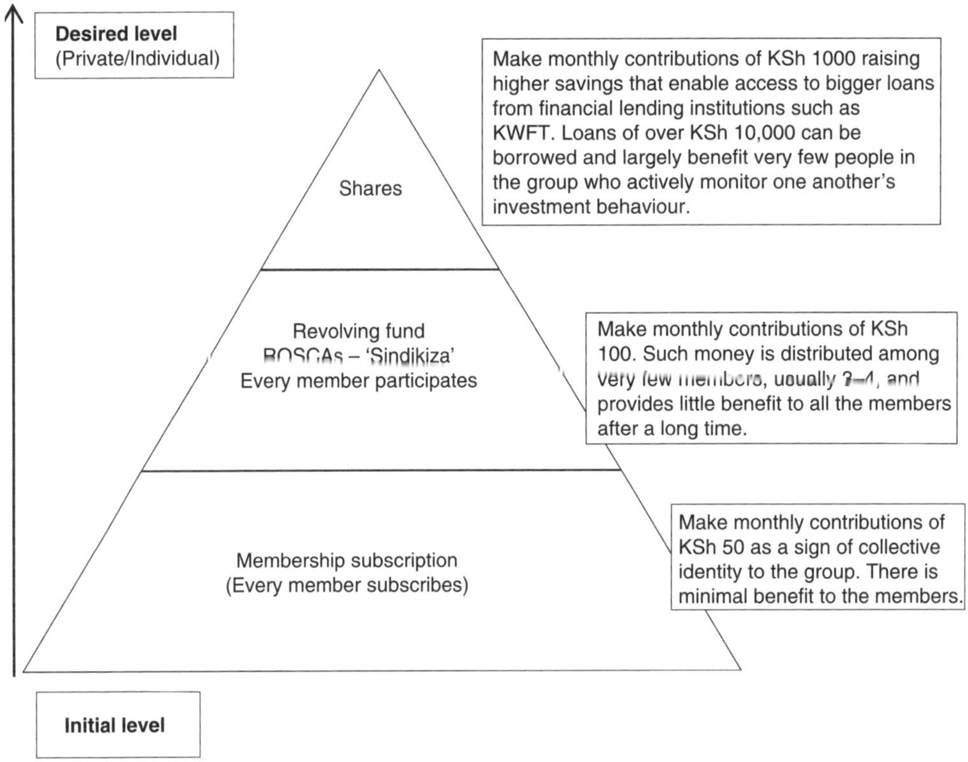


**Fig. 9.4.** Pyramidal benefits from local group participation.

loans for investment (study's focus group discussions).[9] There were a few such groups in all the three districts served by financial institutions.

The interactions among members of a group with a three-tier system of contributions lead to much denser social networks, making them take advantage of the extra benefit the local groups can afford. However, the poorest members may be excluded from the benefits resulting from this level of participation, which are greater than those of lower levels within the economic pyramid.

[9] The rural lending institutions use Grameen Bank model principles of group lending. A loan officer from KREP or KWFT assigned to a division introduces the rural farmers to the credit programme. The potential members self-select into groups, either using the existing groups or forming new ones. The group applies for a single loan, which is divided between members to invest in enterprises of their choice, and the amount given depends on their ability to repay in monthly installments within a period ranging from 2 to 6 months. This arrangement utilizes the potential for peer monitoring and social pressure to enforce compliance in group loan repayment and hence bridges the asymmetric information gap between lenders and borrowers in small-scale credit transactions.

Consequently, one may observe increasingly strong ties between members of substructures within the same community groups and equivalent strength of group pressure against defaulters, which is a key determinant in group performance. This suggests that structural approach to cooperation should examine transfers and exchanges of resources central to the functioning of community groups.

The study findings demonstrate that different types of groups have different capabilities in organizing and leveraging key services and resources to rural households. Three interpretations can be advanced for these differences. First, supra groups may be better resourced to undertake negotiated contractual arrangements with the state and other bodies, and thereby reduce the total costs involved which the local groups cannot afford. Second, supra groups may have a greater effect not only due to their higher liquidity, but also due to an elaborate management and established social responsibility among members. Third, supra groups may enable members to enter into joint investments which can produce greater benefits, hence inducing participation that is more active. Supra groups thus provide both *bridging capital* from heterogeneous groups who work to realize benefits and *linking capital*, which provides the members with opportunities to lobby for support from institutions (Cleaver, 2005).

The study findings raise several implications for future research: To what extent are the community groups able to take on heavy responsibilities and expectations placed on them by communities themselves, policy makers and development practitioners? Do the groups receive the necessary support to meaningfully improve rural livelihoods? Given the financial implications inherent in active memberships, how can the poorest of the poor benefit from group participation that requires meeting upfront costs before realizing benefits? What are the mechanisms for transformation of community groups into supra groups? Given the current heightened attention on community and/or local groups as mechanisms for addressing market and governmental failures, future research into these issues is timely.

## 5 Conclusions and Policy Implications

This chapter has explored patterns of group participation and the impact of group participation on household well-being in Baringo district in Rift Valley Province, Vihiga district in western Kenya and Embu district in eastern Kenya. The results suggest that Embu is the best-endowed district as over 86% of the sampled households have asset indices above the 50th percentile, whereas over 50% of the households in either Vihiga or Baringo are below this baseline. As expected, human, physical and natural capital holdings, as well as gender are important factors influencing household welfare.

But even after controlling for these standard factors, social capital manifest in group participation matters significantly in improving welfare in all three districts. However, community groups, although frequently cited as important for household well-being, have limited effect in this sample. The evidence that such groups are helpful to rural households is mainly anecdotal. On the other hand, membership in supra groups has a statistically significant, positive effect on the households' well-being. Also, services and resources accessed through supra groups generate added statistically significant, positive effects on welfare, after controlling for group membership, suggesting that the larger the range of services and resources offered by supra groups, the greater is the economic benefit enjoyed by members. These groups promote education and training to farmers, improve access to better health and water sources, offer access to financial services, agricultural technology and markets, among other valuable services. To confer these benefits, the power base of groups requires attention by practitioners and policy makers advocating the use of groups in improving welfare and hence reducing poverty.

## 6 Policy Recommendations

Several implications emerge from these findings. First, if social capital accessed through groups is to encourage improvement in the 'missing link' between the function of states and markets in development, the importance of supra groups on economic welfare suggests the need to expand their organizational and resource capacity to benefit more people. This is crucial because, given the differences in capabilities between social affiliation of the poor and the rich, the inequality in group capabilities can be an important source of reducing an individual's well-being as well as creating a potential source of social instability (Stewart, 2005). Group capacity can lead to increased asset accumulation and improved asset productivity, thereby stimulating economic growth and poverty reduction in the local economy.

Second, encouraging the formation of more groups should be checked against their capacity to facilitate access to resources and services such as farm inputs, information, accessing markets and financial services. As hypothesized by Thorp *et al.* (2005), social capital underpinning group formation and maintenance may be important for poverty alleviation, but the chronically poor can be disadvantaged in group formation due to their lesser access to agents capable of politically negotiating their interests. Thus, increasing services through extant groups may be a more desirable intervention than fostering the emergence of new groups, especially community groups. Hence, supporting groups which are not functioning well, assisting them to expand productive opportunities and promoting access to a range of services may be desirable. Due to the likelihood of managing more complex, high-value and costly resources, expanding group organizational structure in terms of leadership, management and community education could help community groups become capable of generating outcomes.

Third, if group participation is to have a durable and equitable effect on rural households, policy makers will need to address the stark disparity across communities and districts in group participation rates and in benefits accessible through the groups. This measure is necessary to fully utilize the potential of groups to improve well-being.

## Acknowledgements

This chapter is drawn from part of the author's dissertation at Cornell University and is made possible in part from sponsorship by the Rockefeller Foundation through the African Food Security and Natural Resource Management (AFSNRM) Program in the Cornell International Institute for Food, Agriculture and Development (CIIFAD). I am greatly indebted to the insightful comments provided by my special committee at Cornell University consisting of Chris Barrett, Margaret Kroma, Lindy Williams and Louise Buck and one member from Egerton University, Abdillahi Aboud. Comments from Andrew Mude, John Omiti and other reviewers are also duly acknowledged. The conclusions presented are mine and do not necessarily reflect the views of the sponsoring organizations.

## References

Akerlof, G.A. and Kranton, R.E. (2000) Economics and identity. *The Quarterly Journal of Economics* 115(3), 715–753.

Amudavi, D.M. (2003) Advancing a partnership model of extension to support the Kenya National Agriculture and Livestock Extension Programme (NALEP) in rural livelihood improvement. Paper presented at the 19th AIAEE Conference, Sheraton Capital Center Hotel, Raleigh, North Carolina, 8–12 April.

Antholt, C.H. (1994) Getting ready for the twenty-first century: technical change and institutional modernization in agriculture. World Bank Technical Paper No. 217.

Barrett, C.B., Reardon, T. and Webb, P. (2001) Nonfarm income diversification and household livelihood strategies in Rural Africa: concepts, dynamics, and policy implications. *Food Policy* 26, 315–331.

Barrientos, A. and Hulme, D. (2005) Chronic poverty and social protection: introduction. *The Journal of Development Research* 17, 1–7.

Bebbington, A. (1999) Capitals and capabilities: framework for analyzing peasant viability, rural livelihoods and poverty. *World Development* 12, 2021–2044.

Berner, E. and Phillips, B. (2005). Left to their own devices? Community self-help between alternative development and neo-liberalism. *Community Development Journal* 40, 17–49.

Byrnes, K.J. (2001) Farmer organizations: tapping their potential as catalysts for change in small-farmer agricultural systems. In: Byrnes, F. (ed.) *Agricultural Extension Systems: An International Perspective.* Erudition Books, North Chelmsford, Massachusetts.

Christoplos, I., Farrington, J. and Kidd, A. (2001) Extension, poverty and vulnerability: inception report of a study for the Neuchatel initiative. Working Paper No. 144. Overseas Development Institute.

Cleaver, F. (2005) The inequality of social capital and the reproduction of chronic poverty. *World Development* 33, 893–906.

Cleaver, K.M. and Schreiber, G.A. (1994) *Reversing the Spiral: The Population, Agriculture, and Environment Nexus in Sub-Saharan Africa.* World Bank, Washington, DC.

Conway, G. (1995) The properties of agroecosystems. *Agricultural Systems* 24, 95–117.

Craig, G. and Mayo, M. (1995) *Community Empowerment: A Reader in Participation and Development.* Zed Books, London.

De Janvry, A., Gordillo, G., Platteau, J. and Sadoulet, E. (2001) Access to land and land policy reforms. In: De Janvry, A., Gordillo, G., Platteau, J. and Sadoulet, E. (eds) *Access to Land, Rural Poverty, and Public Action.* Oxford University Press, Oxford.

De Vaus, D.A. (1991) *Surveys in Social Research*, 3rd edn. Allen & Unwin, North Sydney, New South Wales.

Ellis, F. (2000) *Rural Livelihoods and Diversity in Developing Countries*. Oxford University Press, Oxford.

Filmer, D. and Pritchett, L.H. (2000) Estimating wealth effects without expenditure data-or tear: an application to educational enrollments in states of India. *Demography* 38, 115–132.

Gautam, M. (2000) *Agricultural Extension: The Kenyan Experience, an Impact Evaluation*. World Bank, Washington, DC.

Grootaert, C. (2001) Does social capital help the poor? A synthesis of findings from the local level institutions studies in Bolivia, Burkina Faso and Indonesia. Local Level Institutions Working Paper No. 10. World Bank, Washington, DC.

Grootaert, C. (2004) Social capital, household welfare and poverty in Indonesia. Local Level Institutions Study Working Paper No. 6. World Bank, Social Development Family, Environmentally and Socially Sustainable Development Network, Washington, DC.

Grootaert, C. and Narayan, D. (2004) Local institutions, poverty and household welfare in Bolivia. *World Development* 32, 1179–1198.

Gross, B., van Wijk, C. and Mukherjee, N. (2001) *Linking Sustainability with Demand, Gender and Poverty*. Water and Sanitation Program, Washington, DC.

Heyer, J. Stewart, F. and Thorp, R. (eds) (2002) *Group Behavior and Development: Is the Market Destroying Cooperation?* Oxford University Press, Oxford.

Hill, L.G. and Parker, L.A. (2005) Extension as a delivery system for prevention programming: capacity, barriers, and opportunities. *Journal of Extension* 43(1).

ILCA (1990) *Livestock Systems Research Manual*, Vol. 1. Working Paper 1. International Livestock Centre for Africa, Addis Ababa.

Karega, R.G.M. (1996) Women's groups: from welfare to small-scale business in Kenya. *Small Enterprise Development* 7, 31–41.

Katz, A. (1981) Self-help and mutual aid: an emerging social movement? *Annual Revised Sociology* 7, 129–55.

Kline, P. (1994) *An Easy Guide to Factor Analysis*. Routledge, London.

Knox, A., Meinzen-Dick, R. and Hazell, P. (2002) Property rights, collective action and technologies for natural resource management: a conceptual framework. In: Meinze-Dick, R., Knox, A., Place, F. and Swallow, B. (eds) *Innovation in Natural Resource Management: The Role of Property Rights and Collective Action in Developing Countries*. The Johns Hopkins University Press, Baltimore, Maryland.

Krishna, A. (2002) *Active Social Capital: Tracing the Roots of Development and Democracy*. Columbia University Press, New York.

Krishna, A. (2003) Understanding, measuring and utilizing social capital: clarifying concepts and presenting a field application from India. CAPRi Working Paper No. 28. CGIAR Systemwide Program on Collective Action and Property Rights, Washington, DC.

La Ferrara, E. (2002) Self-help and income generation in the informal settlements of Nairobi. *Journal of African Economies* 11, 61–89.

Lawley, D. and Maxwell, A. (1971) *Factor Analysis as a Statistical Method*. Butterworth, London.

Leana, C.R. and Van Buren III, H.J. (1999) Organizational social capital and employment practices. *Academy of Management Review* 24(3), 538–555.

Lyon, F. (2000) Trust, networks and norms: the creation of social capital in agricultural economies in Ghana. *World Development* 28, 663–681.

Lyon, F. (2003) Community groups and livelihoods in remote rural areas of Ghana: how small-scale farmers sustain collective action. *Community Development Journal* 38, 323–331.

McIntosh, M.K. (2001) Social capital in English communities. In: Rotberg, R.I. (ed.) *Patterns of Social Capital: Stability*

*and Change in Historical Perspective*. Cambridge University Press, Cambridge.
Mknelly, B. and Kevane, M. (2002) Improving design and performance group lending: suggestions from Burkina Faso. *World Development* 30, 2017–2032.
Narayan, D. and Pritchett, L. (1999) Cents and sociability: household income and social capital in rural Tanzania. *Economic Development and Cultural Change* 47, 871–989.
Narayan, D. and Pritchett, L. (2000) Social capital: evidence and implications. In: Dasgupta, P. and Serageldin, I. (eds) *Social Capital: A Multifaceted Perspective*. World Bank, Washington, DC.
Okten, C. and Osili, Una O. (2004) Social networks and credit access in Indonesia. *World Development* 32(7), 1225–1246.
Ostrom, E. (1990) *Governing the Commons: The Evolution of Institutions for Collective Action*. Cambridge University Press, New York.
Place, F., Kariuki, G., Wangila, J., Kristjanson, P., Makauki, A. and Ndubi, J. (2004) Assessing the factors underlying differences in achievements of farmer groups: methodological issues and empirical findings from the highlands of central Kenya. *Agricultural Systems* 82, 257–272.
Platteau, J. and Gaspart, F. (2003) The risk of resource misappropriation in community-driven development. *World Development* 31(10), 1687–1703.
Poole, M.S., Seibold, D.R. and McPhee, R.D. (1996) The structuration of group decisions. In: Hirokawa, R.Y. and Poole, M.S. (eds) *Communication and Group Decision Making*, 2nd edn. Sage Publications, London.
Portes, A. (1998) Social capital: its origins and applications in modern sociology. *Annual Review of Sociology* 24(1), 1–26.
Pritchett, L. and Woolcock, M. (2002) Solution where the solution is *the* problem: arraying the disarray in development. *World Development* 32, 2123–2155.
Prokopy, L.S. (2005) The relationship between participation and project outcomes: evidence from rural water supply projects in India. *World Development* 33, 1801–1819.
Putnam, R.D. (1995) Bowling alone: America's declining social capital. *Journal of Democracy* 6(1), 65–78.
Putnam, R.D. (1996) The strange disappearance of civic America. *American Prospect* Winter, 34–48.
Putnam, R.D. (2000) *Bowling Alone: The Collapse and Revival of American Community*. Simon & Schuster, New York.
Putnam, R.D., Leonardi, I. and Nanetti, R.Y. (1993) *Making Democracy Work: Civic Traditions in Modern Italy*. Princeton University Press, Princeton, New Jersey.
Rees, D., Momanyi, M., Wekundah, J., Ndungu, F., Odondi, J. and Oyure, A.O. (2000) Agricultural knowledge information systems in Kenya – implications for technology dissemination and development. Network Paper No. 107. Agricultural Research and Extension Network, Overseas Development Institute.
Republic of Kenya (1997a) *Embu District Development Plan, 1997–2001*. Government Printer, Nairobi, Kenya.
Republic of Kenya (1997b) *Vihiga District Development Plan, 1997–2001*. Government Printer, Nairobi, Kenya.
Republic of Kenya (1997c) *Baringo District Development Plan, 1997–2001*. Government Printer, Nairobi, Kenya.
Republic of Kenya (2001) *National Agriculture and Livestock Extension Programme (NALEP): Implementation Framework*. Ministry of Agriculture and Rural Development, Nairobi, Kenya.
Robeyns, I. (2005) The capability approach: a theoretical survey. *Journal of Human Development* 6(1).
Sahn, D.E. and Stifel, D. (2003) Exploring alternative measures of welfare in the absence of expenditure data. *Review of Income and Wealth* 49, 4.
Scarborough, V., Killough, S., Johnson, D. and Farrington, J. (1997) *Farmer-led Extension: Concepts and Practices*. Intermediate Technology Publications, Overseas Development Institute, London.

Schouten, T. and Moriarty, P. (2003) *Community Water, Community Management: From System to Service in Rural Areas*. ITDG Publishing, London.

Serageldin, I. and Grootaert, C. (2000) Defining social capital: an integrating view. In: Dasgupta, P. and Serageldin, I. (eds) *Social Capital: A Multifaceted Perspective*. World Bank, Washington, DC.

Stevens, J. (1996) *Applied Multivariate Statistics for the Social Sciences*. Lawrence Erlbaum, Hillsdale, New Jersey.

Stewart, F. (2005) Groups and capabilities. *Journal of Human Development* 6(2).

Stringfellow, R., Coulter, J., Hussain, A., Lucey, T. and McKone, C. (1997) Improving the access of smallholder services in sub-Saharan Africa. *Small Enterprise Development* 8, 35–42.

Thorp, R., Stewart, F. and Heyer, A. (2005) When and how is group formation a route out of chronic poverty? *World Development* 33, 907–920.

Trigilia, C. (2001) Social capital and local development. *European Journal of Social Theory* 4(4), 427–442.

Udvardy, M.L. (1998) Theorizing past and present women's organizations in Kenya. *World Development* 26, 1749–1761.

Uphoff, N., Esman, M.J. and Krishna, A. (1998) *Reasons for Success: Leaning from Instructive Experiences in Rural Development*. Kumarian Press, West Hartford, Connecticut.

Wallace, R. and Wolf, A. (1999) *Contemporary Sociological Theory*, 5th edn. Prentice-Hall, Upper Saddle River, New Jersey.

Wanga, E. (1999) Key note address on new perspectives in rural extension. Regional Refresher International Course in Rural Extension (ICRE) on: Challenges and Prospects. Egerton University, Njoro, Kenya.

Weinberger, K. and Jüting, J. (2000) Women's participation in local organizations: conditions and constraints. *World Development* 29, 1391–1404.

Zeller, M. and Sharma, V. (1998) *Rural Finance and Poverty Alleviation*. Food Policy Report. International Food Policy Research Institute, Washington, DC.

# 10 On Economic Transfers Through Social Networks in Kenya's Smallholder Sector

H. HOGSET

## 1 Introduction

African smallholder farmers have limited access to formal financial markets. Instead, smallholders may turn to informal financial markets. The most important sources of informal finance in rural Africa are private moneylenders and transfers within social networks. The latter usually takes two main forms, either as a collective arrangement often organized as a rotating savings and credit association (ROSCA), in Kenya popularly known as a 'merry-go-round', or as bilateral transfers between individual social network members. This chapter focuses on bilateral transfers within social networks.

Such transfers tend to be rather informal, with no written contracts, and no explicit agreements about the terms of repayment, rather in the spirit of generalized reciprocity. Udry (1994) found that both duration of loans and repayment amounts tended to be state-contingent, with more favourable terms for borrowers who had experienced adverse shocks. Thus, Udry concluded that informal credit could not be clearly distinguished from informal insurance. Indeed, the most important role of this kind of informal finance may be to complement the more rigid ROSCAs in situations of emergency, i.e. to couple informal credit with informal insurance.

Informal transfers do not offer people an opportunity to earn significantly on loans. Indeed, the principle of reciprocity may matter more in the long run than the interpersonal balance of payments. The costs of borrowing through networks are then not so much interest on the loan itself as those associated with the development and maintenance of the network.

A study of informal finance through social networks may inform both further innovation of institutions within the microfinance sector and policies intended to empower the poor and improve their access to financial services of any kind. Institutional design in the microfinance sector is based on assumptions about social mechanisms that govern interactions within social networks

 *Decentralization and the Social Economics of Development* (eds C.B. Barrett, A.G. Mude and J.M. Omiti)

and underpin informal institutions, including borrowing, lending and informal insurance. The key assumptions are that rural villages are socially tight-knit communities in which people know each other well, observe each other's behaviour and where contract enforcement through application of social sanctions is easy. Efficiency of microfinance institutions depends on the validity of these assumptions. However, the literature provides many examples of microfinance schemes that have failed because the social mechanisms the institutional design presumed proved not to function in the environment in which they were introduced (Morduch, 2000; Osterloh and Barrett, Chapter 8, this volume).

But the consequences of such institutional failures are not necessarily evenly distributed across the income distribution. Where resources available for microfinance schemes are limited, it is desirable to design public microfinance programmes that will complement informal finance through social networks, avoid crowding out of beneficial services provided by these extant institutions and avoid exclusion of the most vulnerable from financial services.

## 2 Objective

The dependent variables of this study are frequencies and quantities of material transfers between social network members, whether in cash or in kind. Ordinary market exchanges, i.e. provision of goods or services in exchange for a normal payment, are not considered. This study cannot determine if the market for informal finance through social networks is constrained, be it on the supply side or the demand side, except for a few respondents in given situations. But it can relate bilateral transfers to other sources of finance, including ROSCAs and commercial bank services.

The main objective of this study is to identify suppliers and consumers of this type of finance, and to determine what characterizes the apparent winners and losers in this market, as well as to quantify transfers and derive some general impression of the existence and degree of income pooling and informal insurance against consumption shortfalls among network members. The study also attempts to determine whether social networks are an important source of finance for investments in income-generating activities on- or off-farm.

## 3 Hypotheses

The informal finance market is non-anonymous in the sense that an agent's identity influences market participation. Moreover, network contacts are by nature not instantaneous, but rather relationships that evolve over time and exhibit various degrees of stability and intensity. The frequency and quantities of non-market economic transfers are assumed to be a function of: (i) characteristics of the recipient; (ii) characteristics of the provider; and (iii) characteristics of the relationship between them, including the structure of the social network surrounding their relationship. More specifically, I hypothesize that:

- The probability that two individuals will establish a transfer relationship is a function of the nature of the social tie between them, whether it is conducive to economic transfers – family tie, friendship or other relationship that supports mutual trust and generosity – the physical distance between them and the strategic potential of the relationship, i.e. whether a person is viewed as someone who can provide special benefits to his or her network contacts in the future.
- The quantity of material transfers that an individual is willing to provide to a social network member (an 'alter') depends on his or her capacity to supply transfers and the recipient's ability to reciprocate such favours, both of which are functions of their physical capital, such as size of landholding, and property rights over those assets, their human capital, i.e. education, how their assets are utilized, whether they have off-farm earnings, raise cash crops, etc.; their financial capital, mainly access to formal financial services; and their social position as determined by gender, age and occupation or political role.

A provider of transfers is assumed to go through a two-stage decision process, where one first decides whether to provide anything to the person who requested the transfer, and then decides how much to provide. But a potential provider is only faced with this choice if the two persons involved know each other, and interact beyond some threshold level, i.e. they are members of the same social network. Thus, one may define determination of transfers as a three-stage process, starting with selection of individuals into the same networks. Then they decide whether to form a transfer relationship, and the intensity of this relationship. Finally, for each transfer the level or amount of transfer is decided.

A transfer relationship is assumed to require a mutual desire to maintain such a relationship. A high frequency of interaction within the same dyadic relationship suggests a relationship of some intensity. Lasting transfer relationships are most likely to emerge between individuals who have strong social ties with each other, such as kinship, friendship or a multiplex social relationship, i.e. multiple simultaneous relationships between the same two persons. The degree of social closure within a social environment, or between two individuals, also contributes to strengthening the ties between them, and decreases the social distance between them (Coleman, 1988). But regardless of these contributing factors, the establishment of a lasting transfer relationship is *chosen* (endogenously), not exogenously determined, and unobservable factors are always involved when the choice is made, or when the commitment is renewed.

One of the challenges of the formal financial sector is how to handle asymmetric information. Some recent financial sector innovations targeting the poor seek to exploit the ability of close-knit local communities to overcome such problems through social monitoring of people's behaviour and social enforcement mechanisms that can induce desired behaviour. While the transparency and social control of a village community may be commonly overestimated, a combination of physical proximity and a high degree of social closure reduces the cost of monitoring each other's behaviour, and thereby decreases the incidence of moral hazard, i.e. hidden, unwanted behaviour. Thus, physically distant relations, even those involving family ties, have a handicap that may reduce the probability that a transfer relationship will be maintained, or that will reduce the frequency and quantity of transfers within the relationship.

An individual who holds a high social position that is associated with wealth, power or knowledge is often in a position to offer services that go beyond those ordinary alters can provide. A wealthy person (e.g. a business owner or someone with a salaried occupation) can offer larger material transfers than others, a powerful person (e.g. a civil servant or a village chief) can assist when somebody needs a service from a public office, and an expert of some kind (e.g. a teacher or an extension officer) can give advice on various issues, including problems related to farm production and other income-generating activities. Therefore, people will strategically seek to include such individuals in their social networks, maybe wooing them with services that create a sense of indebtedness, whereby they can establish patron–client relationships for mutual benefits (Platteau, 1995). Such vertical links are known to be important to East Africans. It has been said that in Africa 'the truly destitute is the one who has no patron' (Chabal and Daloz, 1999, p. 42).

Informal insurance can be provided both on terms of specific reciprocity, such as children supporting their elderly parents in return for the services the parents provided the children at a younger age, or in terms of generalized reciprocity, as when those who provide transfers for a funeral do not necessarily expect that the beneficiaries are the same people as the ones who will provide transfers to them when it is their time to have a funeral. This kind of generalized reciprocity may also come into action when villages, for example, collectively raise money to support a student's education. In the case of specific reciprocity, the recipient's ability to reciprocate may be evaluated before the quantity of the transfer is determined, in addition to the recipient's likelihood to repay, when the transfer is considered to be a loan. Thus, observed quantities of transfers are expected to be a function of variables that can explain such abilities. To the extent that these explanatory variables represent permanent features of the beneficiary of transfers, observed transfers can be taken as measures of a person's long-term access to informal finance through social networks.

The expected relationships for standard economic variables such as physical and human capital, wealth and income are relatively straightforward. Individuals who are economically better off should be able to provide larger loans and less likely to default on loans, and should therefore be the providers of larger or more frequent loans, as well as be trusted with larger loans, while they probably need to borrow less often. Being an active lender signals ability to honour loans, and also establishes relationships of patronage, which can be mobilized for raising loans when needed. Thus, socio-economic measures are relevant explanatory variables for the study of both recipients and providers of transfers.

There are reasons to expect that gender might interact with all of the suggested explanatory variables. Studies from Africa have found that males and females tend to establish separate social networks, and that they in particular insure with people of their own gender (M. Goldstein, Yale University, New Haven, 2000, unpublished data). Thus, in an environment where gender interacts with occupational choice and selection into key roles, one should expect the composition of men's and women's social networks to differ with respect to density of persons with access to special resources, such as knowledge, power and wealth. Such differences may be of particular importance to unmarried household heads, if they tend to be excluded from opportunities to tap into networks dominated by the opposite sex.

Age may also interact with these variables, in particular due to life cycle factors. The young poor can be perceived as being temporarily poor, but are expected to improve their income over time, while the elderly poor are more likely perceived as being permanently poor, and thus their consumption emergencies may be perceived as of a more chronic nature. But social networks evolve over time, and the elderly are likely to have richer networks, including younger family members who by local custom have an obligation to support older relatives, especially parents. As with gender, the effect of age on the ability to raise transfers is therefore unclear *ex ante*.

## 4 Econometric Model

The main questions I want my data to answer are: Who engage in material transfers through social networks as recipients or providers? What determines the frequencies and quantities of such transfers? I seek to answer these questions by modelling the decision made by a provider of transfers, which can easily be decomposed into a sequence of two decisions: (i) whether to provide anything at all to a particular person (participation); and (ii) if yes, how much to provide (level). Assuming these two decisions are interlinked and determined by the same explanatory variables, an identification problem may arise. Prior to this decision, the fact that the participants in the transaction interact with each other has been determined by a network selection process that also contributes to the determination of who engages in transactions with whom.

Separate dependent variables are available for the determination of each of these steps: (i) the existence of a network connection; (ii) the frequency of transfers; and (iii) quantity of the largest transfer the preceding year. Existence of a network connection is a binary variable, which takes the value 1 if a connection has been observed, and zero otherwise. The frequency of transfers, i.e. the number of transfers in the last 12 months, is a count variable, dominated by observations of five or less. I assume the decision to participate may be repeated each time a potential lender is being approached by a potential borrower, so the frequency of transfers should be a good measure of a lender's propensity to lend to a particular person, and therefore a better candidate for determining the participation decision than a binary variable that simply takes the value 1 if a transfer has ever occurred, and zero otherwise. The quantity of the largest transfer in the last 12 months is a continuous variable that is left censored at zero. Transfers both in cash and in kind are considered separately. Transfers received and provided are also analysed separately.

These dependent variables necessitate different estimation procedures due to their nature. Estimation of selection into networks requires probit or logit estimation; I use a probit model. Poisson estimations are used to determine the propensity to engage in transfers. Finally, Tobit estimations (conditional on there being positive transfers) are used to determine the quantity of the largest material transfer. This sequence of non-nested regressions represents the baseline for this analysis.

In addition to these, a Heckman model was employed to nest the participation and quantity decisions in one model. The Heckman model estimates the propensity to engage in material transfers first, then in the second (levels) step, controls for selection bias by entering a variable known as Heckman's lambda (Heckman, 1979). The *t* statistic on this variable is taken as a test for the model specification. This test rejected the Heckman estimations for transfers in kind. Therefore, Heckman model results are only reported for cash transfers.

## 5 The Data

The data used in this study were collected during 2003–2004 in two sites in Kenya: Manyatta division in Embu district, and the former Madzuu division (now divided between several new political entities) in Vihiga district, in Eastern and Western Provinces, respectively. Both sites enjoy high, bimodal rainfall patterns, and are considered high-potential areas for agricultural production. Both sites are also extremely densely populated areas. But the ethnicity of the population differs between the two sites, with Embu dominated by the Embu tribe, while the population in Vihiga belongs to the Luhya tribe.[1] These two tribes are culturally different, with different languages and histories, including different histories of interaction with the English during colonial times. Embu also has somewhat more favourable economic conditions than Vihiga, because of generally less-degraded soil, and better market access. Embu is situated about 120 km from Kenya's capital city, Nairobi, while Vihiga is about 400 km from Nairobi, but about 25 km from Kenya's third largest city, Kisumu.

The sample of research subjects was given in both sites, as this study was conducted under the umbrella of a larger interdisciplinary study (BASIS) involving about 120 households in each site. The objective of this larger study is to investigate the dynamic interactions between subsistence agriculture and the natural resource base it depends on, and it involves building a panel data set for the sample households. In both sites a procedure of cluster sampling within multiple villages had been employed. The Vihiga sample was originally surveyed for an unrelated study in 1989. The sample households in Embu were likewise surveyed in 1998 for an unrelated study. Thus, there exists a large data set about the same households which complements the data collected specifically for this study. Data from the BASIS study that have been used in analyses

[1] The concept of a tribe is somewhat politicized in Kenya, and follows only loosely basic identifiers of ethnicity, such as language. Members of the Embu tribe speak a dialect of the Kikuyu language, but insist they are not Kikuyus, while the Luhya tribe is rather a cluster of tribes speaking languages that are related but not mutually intelligible. Yet, they are referred to as one tribe. Thus, one may question what the term 'tribe' means in this context.

presented here were collected in 2002, and they contain information about the households' assets, including farm size and areas under various crops, as well as detailed input and output data for farm operations, a household census, and data on access to credit and how households cope with adverse shocks.

The data collected specifically for this study are a network data set, where the respondents were asked to identify the people they engage with in borrowing and lending (the 'material transfers' network), the people they like to discuss issues of farming with (the 'information transfers' network) and their geographically proximate neighbours. Respondents were asked to list the ten most important network members if the actual number in either one of the categories exceeded ten, but this limit was never binding. These networks also tended to be overlapping. Then, respondents were asked a set of questions about each of these network contacts. The questionnaire uses a 12-month recall period. Within each household, the person who had the main responsibility for day-to-day farm management decisions was selected for interviews. This person is not necessarily the household head. In many households both spouses were farm managers, either working together, or having separate enterprises. Therefore, about one-tenth of the sampled households were selected for interviews with both spouses, where both are farm managers.

The respondents were not asked the same questions about themselves as about alters, so the data about respondents and alters do not mirror each other. Moreover, the rich complementary data set about the sample households that has been collected for the BASIS study does not include alters. Summary statistics are found in the Appendix.

The sample used in regression analysis is the total number of network contacts listed by the respondents. Thus, the control group is realized network relations that are not material transfers relations. A better control group would have been unrealized, but possible, relations, although typically, there are no data available about such a control group. However, I have an extensive data set about a random sample of farmers from 27 different villages, some (but few) of whom listed each other as network contacts. But given that farmers who live in the same village certainly know each other and share similar circumstances, they all could potentially have chosen each other as network contacts. Thus, through a procedure of hypothetical matching, all possible dyads between persons from the same village, whether observed or not, can be used to study endogenous network formation. This was attempted for the network selection regressions that represent step 1 in the baseline, non-nested model, and also in one of the Heckman models estimated as part of the analysis in this study.

Two different matching procedures were chosen, allowing me to use two different sets of data about the matched individuals. First, I pooled network contacts (alters) identified by respondents in the same village. Since respondents were asked about the distance between their own and the alter's residence, distance could be used to exclude alters who were unlikely to have other contacts in ego's village. The set of hypothetical matches actually used in regressions includes alters living 0.5–5.0 km from the respondent who originally named them, thus excluding the potentially least mobile, and also the most distant, persons who were most likely not from the same village. These alters were then matched with all respondents, or egos, from the same village. For this set of matches, the available data include all variables used in the main regressions, except for family ties and distance between residences, which are not known for unrealized dyads. But these hypothetical matches could be pooled with the rest of the data, and included in regressions. These are the ‘ego–alter’ matches. This constructed sample was used in the regressions presented in Tables 10.1 and 10.4.

The second matching procedure consisted of matching respondents with respondents, creating a data set with symmetric information about the matched individuals. But these data do not include the alter information collected for the network study, so this set of hypothetical matches could not be pooled with the rest of the data. These are the ‘ego–ego’ matches. The two matching proced-

**Table 10.1.** Network selection using augmented survey sample. Probit regression.

| | Estimate | | Estimate |
|---|---|---|---|
| *Ego (respondent) characteristics* | | *Alter (network member) characteristics* | |
| Household head | 0.100 | Retired | −0.263 |
| | (0.040)** | | (0.137)* |
| Female | 0.146 | Schoolteacher | −0.271 |
| | (0.039)*** | | (0.064)*** |
| Age | −0.001 | Priest, missionary or | −0.355 |
| | (0.005) | preacher | (0.100)*** |
| ln age | 0.060 | Extension officer, vet, etc. | −0.235 |
| | (0.240) | | (0.063)*** |
| Household head has always lived here | 0.172 | Civil servant or military officer | −0.274 |
| | (0.033)*** | | (0.061)*** |
| Has not completed primary school | 0.072 | Petty trader, shopkeeper | −0.486 |
| | (0.044)* | or business owner | (0.044)*** |
| Has completed secondary school | 0.099 | Village elder or chief | −0.530 |
| | (0.040)** | | (0.118)*** |
| Has completed technical school, | 0.141 | Spokesperson or | −0.380 |
| college or more | (0.076)* | politician | (0.148)** |
| Has house with thatched | −0.378 | | |
| roof (= poor) | (0.090)*** | | |
| Has brick house (= rich) | 0.073 | *Reciprocal characteristics:* | |
| | (0.046) | | |
| Per capita daily total income (US$) | 0.072 | Same sex | 0.525 |
| | (0.022)*** | | (0.030)*** |
| Not recovered from earlier shock | 0.025 | | |
| | (0.035) | | |
| Household has experienced a | 0.095 | Constant | −1.475 |
| health shock last year | (0.040)** | | (0.667)** |
| *Model summary* | | | |
| Number of observations | 10231 | Pseudo $R^2$ | 0.0601 |

Standard errors are given in parentheses.
*Significant at 10%; **significant at 5%; ***significant at 1%.

ures allowed generation of regressions with two different, complementary sets of independent variables that are presented in detail below. This constructed sample was used in the regression presented in Table 10.1.

# 6 Descriptive Statistics

I now turn to a more detailed description of the data. A brief look at some joint frequencies and conditional means reveals some striking relationships that would be lost in regressions unless they were represented by appropriate interaction terms. Inclusion of interaction terms is often desirable, but problematic, due to the inevitable cost of degrees of freedom.

## 6.1 Bilateral transfers

Material transfers through social networks are generally small. In these data, cash transfers range from KSh 10 to 40,000, or about US$0.13–533. (The exchange rate at the time of data collection was about KSh 75/US dollar.) The largest amount came from a network contact living in America. Only two transfers are recorded from providers living more than 500 km away, and none going to someone that distant. Only two cash transfers exceed US$133, both received by respondents in Embu. The data reveal an asymmetry between transfers received and provided, with the former being on average much larger. In Embu the average is US$14.49 receiving and US$8.6 providing, and in Vihiga it is US$6.95 receiving and US$5.21 providing. Dropping the outliers,[2] the average for Embu receiving falls to US$12.25. Men both receive and provide larger amounts than women. The median cash transfer for women is US$2.67, while for men it is US$5.33, both receiving and providing.

Farm inputs constitute about 60% and consumption goods about 30% of all transfers in kind in the data, both receiving and providing. The third most frequent purpose of transfers in kind is ceremonial expenses, which mainly mean funerals. Transfers in kind are typically not evaluated in terms of money, so many respondents were hard-pressed to assign a monetary value to them, but they were asked to suggest an approximate amount they would have been willing to pay for the service provided by the borrowed item or for the goods provided. Obviously, these evaluations were very ad hoc and strongly influenced by unobservable factors, which may be a source of concern in the data.

Transfers in kind in the data range value from US$0.13 to US$666.67, with average transfer in Embu being evaluated at US$3.19 when receiving and US$1.40 when providing, and in Vihiga at US$4.00 when receiving and US$4.24 when providing. (The one extremely large value of a transfer in kind was the value of borrowing somebody's title deed to use it as collateral for a loan

[2] Observations exceeding US$133, i.e. one observation for each of three of the four dependent variables.

meant to bail a person out of prison.) There are two observations greater than or equal to US$133 receiving, both of them from Embu. Dropping them, we get an average value for transfers in kind received of US$1.71 for Embu. Just as with cash transfers, men both receive and provide larger value transfers in kind. The median is US$1.33 for both men and women, but the 90th percentile for men is US$4.87 when receiving and US$2.96 when providing, while for women it is US$2.47 when receiving and US$1.77 when providing. There are no extreme observations on the providing side for either cash or kind.

## 6.2 Social network density and composition

Network density is calculated as the sum of unique names listed by the same respondent as members of their social networks. Mean network density is 11.91 persons in Embu and 8.14 in Vihiga, while it is 10.25 for men and 9.73 for women. Thus, both site and gender influence network density, with the site effect being stronger. The interactions with site and gender are both contrary to expectations. The indigenous perceptions of the cultures in the two sites are that the population in Embu exhibits a greater degree of independence and individuality than that in Vihiga, while women are believed to be more actively networking than men. If the variable 'Household head has always lived here' picks up anything, it should be a continuous history of networking in the same environment, which should result in denser local networks, and indeed mean network density for members of households where this variable is 10.70, while it is 8.93 for others. The variable 'Household head has always lived here' has been chosen as a proxy for general connectedness, but this is clearly a second best, since nearly half (45%) of the respondents are not household heads, so it refers to a different household member.

Household surveys are typically addressed to a household head, who is assumed to be a major decision maker in the household, and be the household member who is best informed about household matters. None the less, the respondents of household surveys are very often not the household heads, and one may wonder what is gained or lost by that. The household census carried out under the BASIS project for this sample identified 98 of my respondents as household heads, 33 as the household head's husband, 87 as the first or second wife and 24 as other or missing. Where the respondent was characterized as the husband, no other household member was assigned the role as household head. In accordance with entrenched perceptions among social scientists, I clustered household heads and husbands together under 'Household head'. But taking it for granted that the identification of a household member as its 'head' is meaningful requires belief in a unitary model of the household, which has faced serious objections from many sources, especially regarding households in sub-Saharan Africa (Alderman *et al.*, 1995).

Traditional economies are typically kinship-based, but the definition of kinship varies between cultures. To identify the closest family ties, my respondents were asked if their network contacts were family members, defined as either a spouse or a next of kin, i.e. a parent, sibling or child. This is a very restrictive definition of kinship that may not be appropriate for the Kenyan context. But to determine what is appropriate, one needs to consider not only the cultural context, but also the relevance of kinship to one's own study, and that has not been done for this study. So, using this definition, 22% of all alters were identified as family members. Family members constitute 14% of the respondents' 'information network' (those they 'like to discuss issues of farming with'), 20% of their neighbours and 28% of the people they engage with in borrowing and lending. The high percentage of family ties with neighbours is surely a result of the customs for land inheritance in the two sites, where land is divided between all the sons when the parents die. The highest concentration of family members is as expected in the transfer networks, but even so, they are still a small minority. Most members of transfer networks are either more distant relations or not relatives at all.

A brief check of the relationship between distance and family ties reveals that 80% of all network contacts living 10 km or more away who provide cash transfers are family members. Remittances from family members not resident in the surveyed households explain, at least partially, the observation that means of transfers are asymmetric, with people receiving more than they provide. This represents a well-known trickle-down of earnings from urban to rural residents, and from the modern to the traditional economy.

Of the three categories of network contacts, the greatest overlap is between neighbours and transfer network members, and the least overlap is between neighbours and the information network. Few respondents include their spouses among the people listed as network members. In some households, both spouses were interviewed. In most of those cases, the spouses did not list each other as network contacts.

Most network connections are between individuals of the same sex, and men's networks are more uniformly same sex than women's. Less than 60% of women's network contacts were the same sex, vs over 80% for men. But both sexes usually list male household heads as neighbours, often even years after their death, perhaps especially when the new household head is female. When network contacts who are included only as neighbours are not counted, over 65% of women's network contacts are the same sex, while the figure for men remains unchanged.

# 7 Regression Results

In this section, results are discussed by variable, organized by category, such as social capital, human capital and physical assets, going across the relevant regressions for selection into transfers, and for frequencies and levels of transfers.

## 7.1 Selection into networks

Results of regressions for selection into networks are presented in Tables 10.1 and 10.2. The dependent variable is a binary variable that takes the value 1 if the combination of ego and alter exists in the network data as a realized network relationship, and zero otherwise. The data used in the regression reported in

**Table 10.2.** Network selection using ego–ego matches. Probit regression.

| | Estimate | | |
|---|---|---|---|
| Constant | –2.270 | | |
| | (0.124)*** | | |
| Both are household heads | 0.430 | | |
| | (0.081)*** | | |
| Same sex | 0.286 | | |
| | (0.080)*** | | |
| Same age (difference < 10 years) | 0.205 | | |
| | (0.079)*** | | |
| Both household heads have always lived here | 0.006 | | |
| | (0.109) | | |
| Neither has completed primary school | 0.027 | | |
| | (0.192) | | |
| Both have completed secondary school or more | 0.347 | | |
| | (0.187)* | | |
| Both have brick houses | 0.587 | | |
| | (0.242)** | | |
| Interaction term non-farm income (US$) | 0.014 | | |
| | (0.012) | | |
| Interaction term farm income (US$) | –0.001 | | |
| | (0.004) | | |
| Both are tea growers | 0.238 | | |
| | (0.128)* | | |
| Both are coffee growers | 0.851 | | |
| | (0.124)*** | | |
| Both raise cattle | 0.107 | | |
| | (0.118) | | |
| Both raise sheep or goats | –0.272 | | |
| | (0.139)** | | |
| Neither has recovered from earlier shocks | 0.167 | | |
| | (0.085)* | | |
| *Model summary* | | | |
| Number of observations | 2340 | Pseudo $R^2$ | 0.1419 |

Standard errors are given in parentheses.
*Significant at 10%; **significant at 5%; ***significant at 1%.

Table 10.1 represent the survey sample augmented with the results of ego–alter matching, while in Table 10.2 they represent the results of ego–ego matching.

The results in Table 10.1 reflect the determinants of network density. The coefficient estimates on both 'Household head' and 'Female' are positive and significant. It may mean that household heads network more actively than other household members and that women network more actively than men. But bear in mind the process of selection into the sample. Respondents were selected among household members who were the main farm managers, which in most households is identical with the household head. Women are underrepresented in the data relative to their proportion of adult household members, and selection of women into the data is a function of characteristics that are not neutral to their propensity to network. So, for both variables there is an obvious endogeneity problem.

Coefficient estimates are positive on all the education variables. The omitted category here is having completed primary school and no more, which represents roughly half of the respondents. The effect is weak in all cases. The poverty proxy, having a house with a thatched roof, has a highly significant and large negative coefficient estimate, suggesting that the poorest are being excluded from social networks. Per capita daily total income has a highly significant and positive coefficient estimate, suggesting that the wealthiest have denser social networks than others. There is also a weak, but significant, positive coefficient estimate on having experienced a health shock last year, suggesting that these households have turned to their networks for help, which increases the reporting of network relations, at least if it means they have been engaged more actively in network transfers than usual.

All of the 'key role' variables, representing alters in positions of high earnings or influence, have strong, negative and highly significant coefficient estimates. They represent cumulatively about 30% of the observed network contacts, but are apparently concentrated in the networks of a few, elite villagers. But the strongest effect of all in this regression is that of belonging to the same sex. A strong, positive effect of belonging to the same sex was expected. A propensity to form gender-specific networks has been found in other studies also (M. Goldstein, Yale University, New Haven, 2000, unpublished data).

For the dyads used in the regression presented in Table 10.2, available data are symmetric between the participants, so it was natural to use it to check which commonalities that are shared by people who choose each other as network contacts, i.e. a qualitative measure of social networks. Thus, the independent variables represent features shared by the two persons in question. Not all of the variables representing shared traits are significant.

The highest and most highly significant coefficient estimates are found on 'Both are household heads', 'Same sex', 'Same age' and 'Both are coffee growers', all of which are positive. The coefficient estimate on 'Both are tea growers' is also positive and significant, but weaker than for coffee. The significant coefficient estimates on gender and headship are probably evidence of network selection, although, as noted above, there is an endogeneity problem associated with these variables. But the significant coefficient estimates on the cash crop variables may simply reflect that these crops are concentrated in the same agroecological zones, so tea growers tend to have other tea growers as neighbours, and likewise for coffee.

There are also significant and positive coefficient estimates on 'Both have secondary school or more' and 'Both have brick houses'. A variable representing 'Both have technical school, college or more' had to be dropped, because no such dyads existed in the data. (Less than 5% of the respondents have completed technical school, college or more.) None the less, the results reveal that people with higher education and people with wealth and high social status seek to network with their equals. Finally, the coefficient estimate is positive and significant on 'Neither has recovered from earlier shocks'. This is a large group – in Vihiga over 80% of the respondents belong to it, so it is quite natural that their networks are dominated by people in the same situation.

To summarize, wealthy and influential people network with others who are also wealthy and influential, while the poorest tend to be excluded. Persons with wealth and influence tend to be concentrated in the networks of a few who presumably belong to the elite in the village. But many traits shared by people in the same networks only reveal that they live in the same neighbourhood, and are not evidence of selection.

### 7.2 Social capital effects

Results of the baseline regressions are presented in Table 10.3. The variables discussed here include social position within the household, i.e. household head, gender, age, and the proxy for general social connectedness 'Household head has always lived here'.

Both household heads and females are more active with transfers in kind, with more frequent transfers both receiving and providing. But household heads provide larger amounts of cash transfers, while females provide cash transfers less frequently, reflecting who in the households have easiest access to cash. The regressions include both a linear and a log linear term for age, and in several regressions non-linear relationships are found. Quantities of cash transfers provided increase up to the age of around 40, and decrease thereafter. This may reflect a peak in labour productivity at that age and therefore in income and the capacity to provide transfers. The frequency of providing transfers in kind also increases with age, up to a maximum at around the age of 50, but levels of transfers in kind are not dependent on age. This pattern suggests a life cycle effect, whereby transfers are passed from the middle-aged to the young.

Households with long histories of networking within the same community, with household heads who have always lived there, are very actively engaged in transfers in kind. They both receive and provide transfers more frequently. This supports the hypothesis that networks evolve and grow denser and more active over time, and is consistent with expectation. However, they receive cash transfers less frequently, and the variable has no effect on levels of transfers. To summarize, the social capital variables are very important for patterns of participation in transfer networks, but not as much for quantities of transfers, especially for transfers in kind.

### 7.3 Human capital effects

The variables discussed here are the three education variables used in the regressions, 'Has not completed primary school', 'Has completed secondary school' and 'Has completed technical school, college or more'. The omitted category is 'Has completed primary school' (the largest group, 56% of the respondents).

**Table 10.3.** Cash transfers and transfers in kind received and provided (converted to US$), using a non-nested model (the baseline).

| | Tobit | Poisson | Tobit | Poisson | Tobit | Poisson | Tobit | Poisson |
|---|---|---|---|---|---|---|---|---|
| | Largest amount received (cash) | Number of times cash transfer received | Largest amount provided (cash) | Number of times cash transfer provided | Largest value received in kind | Number of times transfer in kind received | Largest value provided in kind | Number of times transfer in kind provided |
| Constant | 10.897 | 0.716 | –43.976 | 0.150 | 97.990 | –0.164 | –2.591 | –9.047 |
| | (78.225) | (1.618) | (21.712)** | (1.873) | (102.018) | (1.205) | (9.148) | (1.265)*** |
| Research site | 6.778 | | –1.627 | | 0.632 | | –2.613 | |
| (1 = Embu) | (4.305) | | (2.045) | | (4.171) | | (1.846) | |
| *Ego (respondent) characteristics* | | | | | | | | |
| Household head | 2.472 | –0.067 | 2.789 | –0.122 | –3.045 | 0.287 | 0.880 | 0.238 |
| | (2.921) | (0.084) | (1.551)* | (0.094) | (2.758) | (0.053)*** | (0.555) | (0.048)*** |
| Female | 2.076 | 0.036 | –0.713 | –0.173 | –1.958 | 0.433 | –0.133 | 0.585 |
| | (2.601) | (0.080) | (1.465) | (0.094)* | (1.904) | (0.053)*** | (0.416) | (0.049)*** |
| Age | –0.068 | 0.014 | –0.445 | –0.004 | 0.985 | 0.003 | –0.083 | –0.062 |
| | (0.754) | (0.012) | (0.182)** | (0.014) | (1.026) | (0.009) | (0.077) | (0.009)*** |
| ln age | –2.107 | –0.738 | 17.737 | –0.354 | –37.567 | –0.204 | 1.806 | 2.971 |
| | (29.019) | (0.576) | (7.994)** | (0.666) | (40.175) | (0.419) | (3.176) | (0.437)*** |
| Household head has | –1.943 | –0.171 | –3.426 | –0.118 | 1.503 | 0.484 | 0.881 | 0.182 |
| always lived here | (2.215) | (0.080)** | (2.327) | (0.087) | (1.339) | (0.059)*** | (0.751) | (0.055)*** |
| Has not completed | 4.021 | –0.164 | 1.618 | –0.159 | –3.539 | –0.011 | 1.037 | 0.125 |
| primary school | (4.951) | (0.100) | (1.641) | (0.115) | (4.948) | (0.060) | (0.655) | (0.057)** |
| Has completed | 3.104 | –0.554 | –2.359 | –0.213 | –0.011 | 0.044 | 0.553 | 0.244 |
| secondary school | (3.170) | (0.094)*** | (1.757) | (0.095)** | (1.176) | (0.060) | (0.685) | (0.056)*** |
| Has completed | 15.792 | –0.241 | –2.402 | 0.112 | 1.225 | –0.859 | 1.984 | –0.926 |
| technical school, | (18.895) | (0.186) | (2.057) | (0.166) | (2.836) | (0.129)*** | (2.510) | (0.130)*** |
| college or more | | | | | | | | |

*Continued*

**Table 10.3.** Cash transfers and transfers in kind received and provided (converted to US$), using a non-nested model (the baseline)—*Continued*

| | Tobit | Poisson | Tobit | Poisson | Tobit | Poisson | Tobit | Poisson |
|---|---|---|---|---|---|---|---|---|
| | Largest amount received (cash) | Number of times cash transfer received | Largest amount provided (cash) | Number of times cash transfer provided | Largest value received in kind | Number of times transfer in kind received | Largest value provided in kind | Number of times transfer in kind provided |
| Has house with thatched roof (= poor) | 5.658 | 0.127 | –4.614 | 0.342 | –0.269 | 0.499 | –1.334 | 1.386 |
| | (7.978) | (0.196) | (2.772)* | (0.216) | (2.383) | (0.166)*** | (0.872) | (0.115)*** |
| Has brick house (= rich) | 7.506 | –0.205 | 0.125 | 0.010 | 1.034 | 0.387 | 2.460 | 0.354 |
| | (5.652) | (0.108)* | (2.159) | (0.107) | (1.368) | (0.062)*** | (0.856)*** | (0.060)*** |
| Per capita daily non-farm income (US$) | 15.100 | –0.430 | 11.018 | 0.308 | 0.390 | 0.073 | 0.643 | 0.167 |
| | (4.774)*** | (0.110)*** | (3.651)*** | (0.077)*** | (0.982) | (0.063) | (0.582) | (0.057)*** |
| Per capita daily farm income (US$) | 0.097 | –0.028 | 2.680 | 0.307 | –1.248 | –0.233 | –0.245 | –0.412 |
| | (4.649) | (0.086) | (1.709) | (0.080)*** | (0.996) | (0.049)*** | (0.278) | (0.047)*** |
| Household labour days per month | | 0.001 | | 0.001 | | 0.005 | | 0.001 |
| | | (0.001) | | (0.001) | | (0.001)*** | | (0.001)* |
| Area under cultivation (acres) | | 0.051 | | 0.035 | | 0.095 | | 0.110 |
| | | (0.030)* | | (0.030) | | (0.016)*** | | (0.015)*** |
| ln area under cultivation (acres) | | –0.368 | | –0.074 | | –0.058 | | –0.007 |
| | | (0.070)*** | | (0.078) | | (0.051) | | (0.048) |
| Tea grower | | 0.454 | | 0.304 | | 0.490 | | 0.703 |
| | | (0.075)*** | | (0.077)*** | | (0.043)*** | | (0.041)*** |
| Coffee grower | | –0.032 | | –0.008 | | 0.399 | | 0.418 |
| | | (0.122) | | (0.127) | | (0.084)*** | | (0.080)*** |
| Raises cattle | | 0.736 | | 0.134 | | –0.366 | | –0.289 |
| | | (0.118)*** | | (0.131) | | (0.093)*** | | (0.089)*** |
| Raises sheep or goats | | –0.020 | | –0.203 | | –0.143 | | –0.104 |
| | | (0.080) | | (0.089)** | | (0.050)*** | | (0.048)** |

| | | | | | | | | |
|---|---|---|---|---|---|---|---|---|
| Household receives remittances | −0.575 | −0.128 | −0.712 | 0.465 | −0.980 | 0.077 | 0.186 | 0.035 |
| | (2.120) | (0.065)* | (1.357) | (0.073)*** | (1.167) | (0.044)* | (0.386) | (0.042) |
| A household member has savings account in bank | −0.155 | 0.274 | 2.136 | 0.183 | 2.631 | 0.016 | −0.455 | −0.179 |
| | (3.647) | (0.110)** | (1.867) | (0.120) | (2.674) | (0.063) | (0.575) | (0.060)*** |
| Household head belongs to local savings institution | −3.988 | −0.242 | 0.623 | −0.316 | −0.016 | 0.136 | 0.234 | 0.505 |
| | (4.053) | (0.101)** | (2.306) | (0.109)*** | (1.106) | (0.063)** | (0.418) | (0.063)*** |
| Number of ROSCA memberships | −2.679 | 0.092 | −0.987 | −0.048 | −0.141 | −0.113 | −0.121 | −0.199 |
| | (1.858) | (0.032)*** | (0.689) | (0.035) | (0.529) | (0.023)*** | (0.196) | (0.022)*** |
| Credit constrained in formal credit market | −5.333 | 0.129 | −1.366 | −0.551 | 0.868 | −0.315 | −0.633 | −0.449 |
| | (2.967)* | (0.067)* | (1.191) | (0.074)*** | (1.690) | (0.043)*** | (0.555) | (0.041)*** |
| Credit constrained in informal credit market | 2.483 | −0.277 | 4.861 | −0.459 | −2.027 | −0.931 | −1.219 | −1.482 |
| | (11.309) | (0.194) | (6.895) | (0.219)** | (1.180)* | (0.156)*** | (0.775) | (0.158)*** |
| Household head has not recovered from earlier shock | 3.956 | −0.093 | −0.612 | 0.155 | 2.549 | 0.101 | 0.624 | 0.218 |
| | (5.390) | (0.075) | (1.539) | (0.080)* | (2.191) | (0.045)** | (0.562) | (0.043)*** |
| Household has had a health shock last year | −1.984 | −0.533 | −0.840 | 0.366 | −1.630 | −0.317 | 1.848 | −0.275 |
| | (2.693) | (0.090)*** | (1.087) | (0.081)*** | (1.767) | (0.062)*** | (0.935)** | (0.059)*** |
| *Alter (network member) characteristics* | | | | | | | | |
| Retired | −5.219 | 0.056 | −5.148 | 0.489 | 5.049 | 0.219 | 13.191 | −0.330 |
| | (6.315) | (0.277) | (3.209) | (0.317) | (10.158) | (0.206) | (17.786) | (0.285) |
| Schoolteacher | 5.548 | 0.972 | 3.232 | −0.126 | 1.325 | 0.408 | −0.835 | 0.271 |
| | (4.771) | (0.090)*** | (1.931)* | (0.143) | (1.868) | (0.079)*** | (1.362) | (0.079)*** |
| Priest, missionary or preacher | −3.236 | 0.101 | 1.039 | −0.128 | −3.575 | −0.301 | −1.705 | −0.383 |
| | (6.091) | (0.229) | (2.865) | (0.272) | (2.855) | (0.203) | (1.620) | (0.199)* |
| Extension officer, vet, etc. | 3.434 | −1.878 | 7.137 | −1.599 | −3.362 | −2.231 | −4.324 | −2.434 |
| | (8.704) | (0.411)*** | (4.167)* | (0.320)*** | (5.036) | (0.225)*** | (6.826) | (0.237)*** |

*Continued*

**Table 10.3.** Cash transfers and transfers in kind received and provided (converted to US$), using a non-nested model (the baseline)—*Continued*

| | Tobit | Poisson | Tobit | Poisson | Tobit | Poisson | Tobit | Poisson |
|---|---|---|---|---|---|---|---|---|
| | Largest amount received (cash) | Number of times cash transfer received | Largest amount provided (cash) | Number of times cash transfer provided | Largest value received in kind | Number of times transfer in kind received | Largest value provided in kind | Number of times transfer in kind provided |
| Civil servant or military officer | 10.025<br>(4.726)** | 0.738<br>(0.093)*** | 9.956<br>(5.256)* | –0.818<br>(0.170)*** | 3.701<br>(3.577) | –0.083<br>(0.079) | 2.458<br>(1.763) | –0.388<br>(0.085)*** |
| Petty trader, shopkeeper or business owner | 7.879<br>(3.777)** | 0.283<br>(0.089)*** | 2.899<br>(2.543) | –0.462<br>(0.123)*** | 12.781<br>(13.627) | 0.167<br>(0.067)** | 0.591<br>(1.204) | –0.203<br>(0.072)*** |
| Village elder or chief | 3.734<br>(4.749) | 0.859<br>(0.265)*** | 4.302<br>(2.909) | 0.374<br>(0.309) | –1.999<br>(3.543) | –0.372<br>(0.246) | –0.016<br>(1.033) | –0.394<br>(0.233)* |
| Spokesperson or politician | 1.877<br>(6.276) | –0.481<br>(0.452) | 21.800<br>(39.135) | –0.403<br>(0.452) | –5.308<br>(5.960) | –0.727<br>(0.230)*** | 1.145<br>(0.780) | –1.226<br>(0.335)*** |
| *Reciprocal characteristics* | | | | | | | | |
| Same sex | –5.485<br>(3.797) | 0.149<br>(0.062)** | 0.763<br>(1.180) | 0.748<br>(0.080)*** | –0.472<br>(1.220) | 0.126<br>(0.041)*** | 0.180<br>(0.357) | 0.137<br>(0.038)*** |
| Close family member | 3.670<br>(5.815) | 1.133<br>(0.056)*** | 0.153<br>(1.312) | 0.653<br>(0.065)*** | 0.174<br>(1.467) | 0.298<br>(0.042)*** | –0.114<br>(0.496) | 0.564<br>(0.038)*** |
| Distance between residences (km) | 0.012<br>(0.017) | 0.000<br>(0.000)*** | 0.298<br>(0.880) | –0.040<br>(0.017)** | –0.000<br>(0.032) | 0.000<br>(0.000) | 0.461<br>(0.251)* | –0.009<br>(0.003)*** |
| *Purpose of largest transfer* | | | | | | | | |
| Farm investment | 24.136<br>(11.973)** | | 5.348<br>(3.113)* | | 11.101<br>(10.669) | | 3.903<br>(1.905)** | |
| Farm inputs | 3.940<br>(4.368) | | 1.426<br>(2.093) | | –0.451<br>(2.675) | | 1.008<br>(1.141) | |

| | | | | | | | | |
|---|---|---|---|---|---|---|---|---|
| Off-farm business expenses | 43.169 | | 27.654 | | –1.141 | | 2.402 | |
| | (16.004)*** | | (10.581)*** | | (4.466) | | (1.232)* | |
| School fees | 28.106 | | 9.068 | | 4.912 | | 7.969 | |
| | (12.193)** | | (2.686)*** | | (3.916) | | (3.162)** | |
| Medical expenses | 4.549 | | –0.538 | | 1.476 | | 0.567 | |
| | (4.087) | | (1.407) | | (3.561) | | (0.643) | |
| Ceremonial expenses | –2.447 | | 2.837 | | 4.080 | | 8.003 | |
| | (3.345) | | (4.054) | | (6.166) | | (6.396) | |
| *Model summary* | | | | | | | | |
| Number of observations | 481 | 2046 | 434 | 2046 | 723 | 2046 | 737 | 2046 |
| Pseudo $R^2$ | 0.0704 | 0.1557 | 0.0710 | 0.1095 | 0.0048 | 0.1200 | 0.1562 | 0.1764 |
| *P* value White's test | 0.0289 | | 0.0696 | | 0.0000 | | 0.0000 | |

Standard errors are given in parentheses. Jackknife standard errors for Tobit regressions.
*Significant at 10%; **significant at 5%; ***significant at 1%.

Of the respondents, 16% have not completed primary school, while 4.6% have completed technical school or college.

Those with no formal education provide transfers in kind more frequently, but this effect is weak. In comparison, those who have completed technical school, college or more both receive and provide transfers in kind less frequently, and this effect is strong and highly significant. Those with secondary school engage in cash transfers much less frequently, both receiving and providing, while they provide transfers in kind more frequently.

This pattern is a little unclear, but a suggested structure is that those with no formal education are poor, those with secondary school represent an intermediate wealth level, while those with technical school or college are the wealthiest people in the sample. Apparently, the poorest engage mostly in frequent, but low-value transfers in kind; those of intermediate wealth are actively engaged in transfers in kind but avoid cash transfers, maybe to protect meagre cash earnings; and the richest do not engage much in transfers through networks at all, because they have other, preferred options.

## 7.4 Wealth, income and household assets

Here, I will first discuss variables representing wealth and income, and then turn to variables that represent household assets. The variables discussed first are 'Thatched roof', 'Brick house' and per capita farm and non-farm income (in US$). 'Thatched roof' serves as a proxy for poverty and 'Brick house' as a proxy for wealth and social status. Traditionally, people in both sites did not consider a home to be a permanent structure to be passed down through generations, but rather a personal item, maintained during the owner's lifetime, and then left to the elements to crumble. The traditional building material is mud, fortified with wooden sticks, and in Vihiga, most sample households (82%) have mud houses. In Embu, wood structures have become common, and 70% of the sample households there live in wood houses. But brick houses are considered the best in both places, and represent a status symbol. In Embu, 11% of the respondents live in brick houses vs 15% in Vihiga. Thatch is the traditional roofing material, and has become uncommon in both sites. Only about 3% of the respondents live in houses with thatched roofs, and they presumably represent the poorest among them. Most residential houses, including the fancy brick houses, have corrugated tin roofs.

For cash transfers, the coefficient estimates on 'thatched roof' are weak – only one is mildly significant, and may represent a spurious correlation due to the very low number of non-zero observations. However, residents in houses with thatched roofs engage more actively in transfers in kind, though the variable has a weak (insignificant), negative effect on levels of transfers. Residents in brick houses receive cash transfers less frequently, but receive and provide transfers in kind more frequently. They also provide higher-value transfers in kind.

The coefficient estimate on non-farm income is positive, highly significant and dwarfs most other effects for levels of cash transfers, both received and provided. A high non-farm income increases the frequency of providing transfers, both in cash and kind, while it decreases the frequency of receiving cash transfers, indicating that these persons do not borrow frequently, but are themselves sources of credit for others. In contrast, farm income increases the frequency of providing cash transfers, while it decreases frequency of transfers in kind both ways, but it has no effect on levels of transfers.

The remaining variables in this category represent household assets, and were only used in regressions for frequencies, and not for levels of transfers. The variable 'Household labour days per month' is the sum of labour days per month provided to the family farm by household members at least 10 years of age. This variable has no effect on cash transfers, but is associated with more frequent transfers in kind, both receiving and providing. This may mean that large households are very actively engaged in network transfers. The coefficient estimates on holding size (linear and log-linear terms to capture non-linear relationships) are significant for frequencies of cash transfers received, and is diminishing in holding size within the relevant size range. This may be a wealth effect – the richer have access to other sources of finance, and are not so active in transfer networks. The effects of holding size on frequencies of transfers in kind are weak, but positive and highly significant.

There is a very strong, positive effect of cash crops (both tea and coffee) on the frequency of transfers in kind both ways. Tea growers also engage more frequently in cash transfers, both ways. In contrast, the coefficient estimates for the livestock variables 'Raises cattle' and 'Raises sheep or goats' are negative for frequencies of transfers in kind both ways, suggesting that livestock owners tend to withdraw from transfers in kind. Cattle owners receive more frequent cash transfers, while owners of small ruminants provide cash transfers less frequently. The coefficient estimates on these variables are difficult to interpret, and may be due to correlation with unobserved factors, such as the ability to use livestock as collateral, or possibly a role of livestock as instruments of saving.

The important relationships described here relate to wealth and endowment effects on transfers. The wealthy, those with large farms, high non-farm incomes and brick houses, receive less frequent, but higher, amounts of cash transfers, but provide cash transfers relatively frequently. Farmers with cash crops (tea) are more likely to provide cash transfers, and cash transfers are larger if coming from persons with a high non-farm income. A high farm income decreases participation in transfers in kind, but non-farm income and a high status (brick house) increase it.

## 7.5 Access to formal and semi-formal finance and insurance

This section explores the effects of variables related to formal and semi-formal sources of finance, shocks and insurance. These variables include membership in ROSCAs, being a recipient of remittances, having a household member with a savings account in a bank and/or who is a member in a local savings institution, and being credit-constrained in the formal and informal credit markets. 'Remittances' refers to household cash incomes earned by household member(s) living elsewhere. This is not equivalent to cash incomes earned by a household resident, as these payments are more likely to arrive when the household is in economic distress. Thus, remittances may be considered a form of informal insurance. I will also discuss variables that represent risk and vulnerability.

Effects of these variables on quantities of transfers are almost totally absent, but there are some strong relationships on frequencies of transfers. Members of households that receive remittances receive cash transfers through social networks less frequently than others, but they provide cash transfers more frequently, suggesting some pass-through or trickle-down effect of remittances. This is the same pattern as exhibited by the rich, who do not need bilateral transfers for credit, but are able to provide credit to others. The remittances variable is associated with receiving more frequent transfers in kind, which may represent how cash transfers provided are reciprocated.

The coefficient estimates on access to formal finance are inconsistent and difficult to make sense of. It may be because of the way formal finance interacts with other variables. Those who have either a savings account in a bank or are a member of a local savings institution (SACCO) are all from Embu, and they all own cattle. A majority of both tea and coffee growers in Embu either have bank accounts or are members of a SACCO (70–80%). Thus, they have many characteristics of the wealthy. But they are not more likely than others to have a brick house, the status symbol used as a proxy for wealth here. Those who have savings accounts in banks receive cash transfers more frequently and provide transfers in kind less frequently, while those who are members of SACCOs both receive and provide less cash and both receive and provide more in kind. Apparently, bank accounts are associated with greater wealth than membership in SACCOs. Membership in ROSCAs is associated with less participation in transfers in kind, and with receiving cash transfers more frequently.

Being credit-constrained in either the formal or the informal credit market reduces the frequencies of transfers in kind both ways significantly, and it also reduces frequencies of cash transfers provided, reflecting their poor access to cash and therefore inability to provide transfers. However, those who are credit-constrained in the formal credit market do receive cash transfers more frequently, albeit at lower levels. In contrast, those who are credit-constrained in the informal credit market receive and provide fewer transfers of both kinds, and receive lower quantities of transfers in kind. These may be persons who are not trusted among the villagers, and therefore are cut off from informal finance, while those who are credit-constrained in the formal market may be hampered by lack of collateral, a problem that it may be possible to circumvent in the informal credit market, if the person is otherwise considered trustworthy.

The 'earlier shock' variable is positively associated with frequencies of transfers in kind, both receiving and providing, suggesting that having suffered a shock increases people's dependency on social networks, although the coefficients are small. This variable is also associated with a higher frequency of cash transfers provided, but not received, which is contrary to expectation.

Perhaps it is evidence that the vulnerable invest in social networks as a source of social insurance.

The 'health shock' variable is associated with a lower frequency of cash transfers received, and lower frequencies of transfers in kind both ways. This may be evidence that health shocks make people drop out of their social networks, or some form of exclusion, whether intentionally or not. The strong negative effect on cash transfers received is disturbing. It suggests that social networks do not provide informal insurance for the seriously ill. And, perversely, households with a severely ill member provide more cash transfers, a pattern that is reminiscent of the transfer pattern for households that have not recovered from earlier shocks – the more difficult your situation, the more you provide, and the less you receive, apparently.

## 7.6 Social network contacts and relationships

The 'occupation' variables are answers to the question 'is this person (i.e. alter) a key person in this village or your life, like ...' followed by examples, including village chief, schoolteacher, etc. I hoped this formulation would ensure that respondents would not consider only the person's main source of income, but also other roles, like being an elder, which may be associated with social status, but not necessarily income. Most network contacts were thus characterized as 'none of the above', because they were only ordinary farmers. More than 70% of the network contacts were 'none of the above'. The most common key roles in the data are schoolteacher, extension officer, vet or representative of an agricultural cooperative, civil servant or military officer, and petty trader, shopkeeper or business owner. Others also included in regressions are priest, missionary or preacher, elder or chief, and politician or spokesperson. There is also a variable representing an alter who has retired from a position coded as a 'key role'. (There are no significant coefficient estimates on this variable, so it will not be commented on further.) Note that when the respondent is a provider of transfers, alter is the recipient, and vice versa. The terms 'received' and 'provided' in the column headings of the tables refer to the respondent, so the opposite is true for alter.

The names of extension officers typically came as answers to the question 'Who do you like to discuss issues of farming with?', which indicates that extension officers belong to information networks. Some of them were listed by many. Up to 15–20 different respondents could list the same extension officer (nobody who was not an extension officer was ever listed by more than six respondents). Few of these reported any borrowing or lending with them. As a result, coefficients on extension officers in the material transfer regressions are consistently negative. This should not be taken as evidence of extension officers' behaviour as members of material transfer networks – they were just not members of my respondents' material transfer networks.

In rural villages in Kenya, the village schoolteacher is among the few with salaried employment, making them among the least cash-constrained residents there. As a consequence, they are pulled into transfer networks. Schoolteachers engage in transfers in kind more frequently, both receiving and providing. Respondents receive cash transfers more frequently from teachers, while they provide higher amounts of cash transfers to them, suggesting that schoolteachers belong to the relatively wealthy.

There are fewer civil servants and military officers in the villages, but they are privileged not only by having high and steady cash incomes, but also because of their access to power. Thus, people woo them as network contacts, and invest in network relationships with them as useful contacts to have whenever they need to deal with government offices. Conversely, civil servants may be interested in exploiting their position to elicit favours. And as expected, civil servants are highly visible members of these social networks. Respondents receive more frequent and higher-level cash transfers from them, while they provide less frequent, but higher-level cash transfers to them. However, respondents provide transfers in kind less frequently to civil servants and military officers. This is the pattern of transfers with the wealthiest.

Many rural residents engage in petty trade to augment their farm incomes, and if they succeed with that, they should be less cash-constrained than the average subsistence farmer. However, when making purchases from petty traders, one quickly realizes that the cash amounts that pass through their hands are very small – they usually have great difficulties giving back the balance on even the smallest notes. But respondents did not distinguish between these petty traders and persons owning larger, more professional business operations, so this category is very diverse. It includes shopkeepers who let farmers buy farm inputs on credit. The coefficients on this variable are similar to those of civil servants, only with smaller effects. Respondents are less likely to provide transfers to shopkeepers, whether in cash or kind, but they receive both types of transfers from them more frequently. These transfers are probably loans rather than gifts, and are therefore more akin to commercial lending than most transfers within social networks.

Priests, missionaries and preachers, and politicians and spokespersons are hardly visible at all in the regressions. Both of the study sites are homes of powerful politicians. One former vice president comes from the same neighbourhood as where the Vihiga data were collected, while a cabinet minister at the time of data collection comes from the study site in Embu. Neither of these shows up in the data, but politicians who do appear in these networks probably have connections to these national figures, or are able to call upon them in special circumstances. However, they do not show up in these regressions as active members of transfer networks.

There is a strong, positive coefficient estimate on 'Village chief or elder' in the regression for frequency of cash transfers that respondents receive. Some of these persons are elderly men with a past in high-ranking and well-paid positions, but it is by no means a rule that they are wealthy. However, some respondents are able to borrow cash from such persons.

In general, those who have been able to include these key persons in their networks enjoy access to frequent cash loans, which must be very useful in this cash-strapped economy. Apparently, such key persons are concentrated in the social networks of a lucky few.

## 7.7 Reciprocal characteristics

Three variables are included in the regressions to represent the relationship between respondent and alter, the 'reciprocal' characteristics: 'Same sex', 'Close family members' and 'Distance between residences (km)'. Just as in the network selection regressions discussed above, coefficient estimates on the variable for same sex are consistently strong and positive in these regressions, at least for frequencies of transfers. The same is true for the family member variable. But neither of these variables has a significant effect on levels of transfers. This means persons of the same sex may be networking intensely with each other, but they are not sources of major transfers. This is probably particularly true for females, whose female network contacts tend to be as cash-strapped as they are themselves.

In many studies of informal and semi-formal finance in rural communities, material transfers through social networks are assumed to be clan- or kinship-based systems of material support, but as mentioned above, in my data family members constitute less than one-third of all transfer network members. One important difference, though, is the very restrictive definition of family ties that has been applied here. It is both possible and likely that a majority of social network members are members of the respondents' own clan, while not being a close family member. The clan represents a less cohesive income-pooling unit than a nuclear family, but being much larger, it may exhibit a lower degree of covariate risk, and therefore have a higher ability to insure against idiosyncratic risk.

As discussed earlier, distance is related to remittances. More geographically distant network contacts are obviously not neighbours, and rarely listed as people the respondents like to discuss issues of farming with. They were mentioned because of the remittances they provide. A few respondents mentioned alters living as far away as America, which makes them outliers with respect to distance. This turns up in the regression results for frequencies of transfers – here, coefficient estimates on distance are near zero, yet significant. Remittances are typically asymmetric, and represent a one-way flow of cash going from the urban to the rural economy, as well as between generations within families.

## 7.8 Purposes of transfers

Purposes of transfers were only included in the Tobit models and levels stage in the Heckman models because purposes were only observed for non-zero transfers. The excluded purposes are consumption, which is the most common purpose for transfers both in cash and in kind, and investment in a residential building. The latter was excluded from regressions because it appears only sporadically in the data, and had to be excluded from most regressions anyway. Three purpose variables stand out as consistently strong and positive. Those are farm investments, investments in off-farm businesses and school fees.

All the coefficients on medical and ceremonial expenses are insignificant. This is contrary to expectations, given the ongoing AIDS crisis in both study sites, where it was often difficult to find any respondents, because they were too busy attending funerals. Maybe a reason is a lack of access to medicines, so health shocks do not generate cash demands. It is also possible that the excluded category 'consumption' includes many transfers that were intended to assist with consumption smoothing through periods of illness. But if the respondent expressed that the transfer was needed due to illness in the household, it should have been recorded as a medical expense.

### 7.9 The Heckman models for cash transfers

Results of the Heckman model estimations are presented in Table 10.4. Significance of the models themselves can be judged by the $P$ value of Heckman's lambda. For all four regressions this test statistic is significant at the 10% level or higher. (Only regressions representing cash transfers are included.) Two different samples were used – the same sample as for the baseline regressions ('survey sample'), and the augmented sample also used to estimate network selection (Table 10.1). The augmented sample includes hypothetical ego–alter matches within villages.

There are very few cases of contradictions with the baseline. Only one variable has a highly significant coefficient estimate of the opposite sign as in the baseline. That is the variable 'Petty trader, shopkeeper or business owner', which has a negative effect on selection into cash transfers received here, but only in the regression using the augmented sample. This means the probability that a shopkeeper is selected into cash transfer networks is low, and this effect cancels the probability of extending cash loans, given membership in such a network, which is high.

The coefficient estimates on some variables are significant in the Heckman models for both samples, although they were insignificant previously. Two are found in the regressions for cash transfers provided. Ego is more likely to provide cash transfers if he or she has completed technical school, college or more, and if one lives in a brick house. This supports the hypothesis that the wealthier provide more cash transfers. In all four regressions, the same sex variable has a significant effect on levels of cash transfers. This effect is positive for transfers provided, but negative for transfers received. This asymmetry is difficult to explain.

In the Heckman regressions, the outliers regarding distance between ego and alter were dropped. Thus, coefficient estimates on the distance variable are more informative. For cash transfers received, distance has a positive effect on both selection into, and levels of, transfers, while for transfers provided, distance has no effect. This reflects the asymmetric flow of remittances into the rural households.

## 8 Conclusions

One of the most important findings of this study is that resources in terms of human capital and the stock of productive assets play an important role in individuals' selection into networks. There is a tendency for people to network

**Table 10.4.** Cash transfers received and provided (converted to US$), using a two-step Heckman selection model. The sample is either the original survey sample or this sample augmented with hypothetical ego–alter matches (see text).

| | Survey sample | | Augmented sample | | Survey sample | | Augmented sample | |
|---|---|---|---|---|---|---|---|---|
| | Cash received | | Cash received | | Cash provided | | Cash provided | |
| Step | (1) | (2) | (1) | (2) | (1) | (2) | (1) | (2) |
| | Largest amount received | Selection | Largest amount provided | Selection | Largest amount provided | Selection | Largest amount provided | Selection |
| Constant | 13.784 | 1.676 | 58.587 | −2.128 | −18.704 | 3.144 | −35.781 | −1.015 |
| | (9.540)** | (1.900) | (29.227)** | (0.153)*** | (9.314)** | (1.927) | (11.954)*** | (1.071) |
| Research site | 6.130 | | | | −2.728 | | −5.274 | |
| (1 = Embu) | (7.908) | | | | (4.396) | | (4.178) | |
| *Ego (respondent) characteristics* | | | | | | | | |
| Household head | 2.887 | −0.072 | −1.666 | 0.001 | 0.975 | −0.054 | 2.619 | 0.037 |
| | (3.492) | (0.098) | (4.086) | (0.066) | (2.390) | (0.100) | (2.062) | (0.070) |
| Female | 0.558 | −0.078 | −3.116 | 0.032 | −5.301 | −0.137 | −2.623 | 0.017 |
| | (3.271) | (0.093) | (3.865) | (0.062) | (2.261)** | (0.095) | (1.855) | (0.065) |
| Age | −0.143 | 0.011 | −0.000 | −0.003 | −0.293 | 0.014 | −0.150 | −0.000 |
| | (0.098) | (0.014) | (0.107) | (0.002) | (0.097)*** | (0.014) | (0.064)** | (0.009) |
| ln age | | −0.877 | | | | −1.302 | | −0.314 |
| | | (0.675) | | | | (0.686)* | | (0.393) |
| Household head has always lived here | −3.420 | 0.119 | −0.938 | 0.034 | −1.554 | 0.072 | −1.155 | −0.016 |
| | (3.371) | (0.088) | (3.574) | (0.058) | (2.350) | (0.090) | (1.982) | (0.061) |
| Has not completed primary school | 5.873 | −0.018 | 4.965 | −0.088 | 2.900 | 0.154 | 0.171 | 0.007 |
| | (3.532)* | (0.101) | (4.227) | (0.072) | (2.568) | (0.103) | (2.176) | (0.078) |
| Has completed secondary school | 2.544 | −0.167 | 1.623 | −0.071 | −1.844 | 0.059 | −1.318 | 0.057 |
| | (3.115) | (0.091)* | (3.637) | (0.062) | (2.086) | (0.090) | (1.795) | (0.064) |

*Continued*

**Table 10.4.** Cash transfers received and provided (converted to US$), using a two-step Heckman selection model. The sample is either the original survey sample or this sample augmented with hypothetical ego–alter matches (see text)—*Continued*

| | Survey sample | | Augmented sample | | Survey sample | | Augmented sample | |
|---|---|---|---|---|---|---|---|---|
| | Cash received | | Cash received | | Cash provided | | Cash provided | |
| Step | (1) | (2) | (1) | (2) | (1) | (2) | (1) | (2) |
| | Largest amount received | Selection | Largest amount provided | Selection | Largest amount provided | Selection | Largest amount provided | Selection |
| Has completed technical | 8.699 | 0.215 | 20.052 | 0.171 | 2.288 | 0.416 | 1.653 | 0.299 |
| school, college or more | (6.484) | (0.177) | (6.931)*** | (0.124) | (4.354) | (0.17?)** | (3.375) | (0.116)*** |
| Has house with | 2.602 | 0.287 | 8.429 | –0.080 | 3.056 | 0.456 | –2.778 | 0.054 |
| thatched roof (= poor) | (7.291) | (0.218) | (8.715) | (0.135) | (5.360) | (0.220)** | (3.955) | (0.136) |
| Has brick house (= rich) | 6.111 | 0.112 | 7.003 | 0.116 | 4.397 | 0.244 | 3.753 | 0.210 |
| | (3.859) | (0.111) | (4.380) | (0.077) | (2.725) | (0.109)** | (2.256)* | (0.075)*** |
| Per capita daily non- | 15.305 | –0.127 | | | 13.119 | 0.078 | 13.058 | 0.072 |
| farm income (US$) | (2.958)*** | (0.087) | | | (1.858)*** | (0.079) | (1.575)*** | (0.058) |
| Per capita daily farm | –0.372 | –0.125 | | | –0.763 | –0.140 | 0.218 | –0.110 |
| income (US$) | (3.229) | (0.086) | | | (1.991) | (0.083)* | (1.631) | (0.060)* |
| Household daily non- | | | 3.364 | –0.026 | | | | |
| farm income (US$) | | | (0.876)*** | (0.014)* | | | | |
| Household daily farm | | | 0.696 | –0.018 | | | | |
| income (US$) | | | (0.580) | (0.013) | | | | |
| Household size | | | –1.549 | 0.034 | | | | |
| | | | (0.727)** | (0.013)*** | | | | |
| Household labour | –0.048 | 0.002 | | –0.001 | 0.010 | 0.002 | –0.014 | –0.000 |
| days per month | (0.045) | (0.001)* | | (0.001) | (0.032) | (0.001) | (0.028) | (0.001) |
| Area under cultivation | 2.089 | –0.021 | | –0.007 | 0.641 | 0.020 | 0.616 | 0.035 |
| (acres) | (1.009)** | (0.035) | | (0.028) | (0.468) | (0.030) | (0.642) | (0.023) |
| ln area under | | –0.077 | | –0.071 | | –0.018 | 0.296 | –0.056 |
| cultivation (acres) | | (0.083) | | (0.057) | | (0.080) | (1.608) | (0.056) |

| | | | | | | | | |
|---|---|---|---|---|---|---|---|---|
| Tea grower | 0.379 | 0.141 | | 0.151 | 4.245 | 0.098 | 5.035 | 0.112 |
| | (2.979) | (0.081)* | | (0.057)*** | (2.042)** | (0.082) | (1.753)*** | (0.060)* |
| Coffee grower | –2.453 | 0.115 | | 0.376 | 3.436 | 0.024 | 9.064 | 0.378 |
| | (7.053) | (0.111) | | (0.077)*** | (4.039) | (0.113) | (3.897)** | (0.079)*** |
| Household head has not recovered from an earlier shock | 3.556 | –0.010 | –0.282 | 0.029 | –1.189 | –0.032 | –0.873 | –0.009 |
| | (2.743) | (0.079) | (3.199) | (0.056) | (1.873) | (0.079) | (1.614) | (0.056) |
| Household has had a health shock last year | –[illegible].190 | –0.183 | –2.780 | –0.005 | –1.872 | –0.039 | –0.458 | 0.061 |
| | (3.363) | (0.092)** | (3.715) | (0.063) | (2.132) | (0.092) | (1.842) | (0.065) |
| *Alter (network member) characteristics* | | | | | | | | |
| Retired | –5.875 | 0.371 | –4.386 | –0.004 | 1.035 | 0.504 | –7.046 | 0.012 |
| | (8.769) | (0.304) | (9.673) | (0.167) | (7.002) | (0.320) | (5.784) | (0.203) |
| Schoolteacher | | 0.673 | –1.408 | 0.226 | | –0.142 | | –0.167 |
| | | (0.137)*** | (5.058) | (0.082)*** | | (0.154) | | (0.103) |
| Priest, missionary or preacher | | 0.207 | –2.775 | –0.054 | | 0.002 | | –0.161 |
| | | (0.240) | (9.315) | (0.148) | | (0.263) | | (0.164) |
| Extension officer, vet, etc. | | –0.705 | 9.684 | –0.693 | | –0.863 | | –0.783 |
| | | (0.215)*** | (13.203) | (0.153)*** | | (0.228)*** | | (0.163)*** |
| Civil servant or military officer | | 0.322 | 4.382 | 0.063 | | –0.204 | | –0.273 |
| | | (0.130)** | (4.796) | (0.082) | | (0.143) | | (0.100)*** |
| Petty trader, shopkeeper or business owner | | 0.247 | 8.823 | –0.154 | | –0.233 | | –0.403 |
| | | (0.108)** | (4.378)** | (0.067)** | | (0.120)* | | (0.079)*** |
| Village elder or chief | | 0.600 | 2.810 | –0.175 | | 0.145 | | –0.314 |
| | | (0.296)** | (11.167) | (0.174) | | (0.319) | | (0.199) |
| Spokesperson or politician | | –0.018 | –4.744 | –0.248 | | –0.534 | | –0.624 |
| | | (0.362) | (14.679) | (0.236) | | (0.421) | | (0.306)** |
| *Reciprocal characteristics* | | | | | | | | |
| Same sex | –7.322 | 0.293 | –14.558 | 0.521 | 6.684 | 0.342 | 9.262 | 0.539 |
| | (2.765)*** | (0.074)*** | (5.986)** | (0.048)*** | (2.521)*** | (0.076)*** | (2.686)*** | (0.051)*** |
| Close family member | 0.951 | 0.529 | 5.078 | | 8.750 | 0.509 | 1.023 | |
| | (3.404) | (0.074)*** | (2.687)* | | (3.018)*** | (0.075)*** | (1.234) | |

*Continued*

**Table 10.4.** Cash transfers received and provided (converted to US$), using a two-step Heckman selection model. The sample is either the original survey sample or this sample augmented with hypothetical ego–alter matches (see text)—*Continued*

| | Survey sample | | Augmented sample | | Survey sample | | Augmented sample | |
|---|---|---|---|---|---|---|---|---|
| | Cash received | | Cash received | | Cash provided | | Cash provided | |
| Step | (1) | (2) | (1) | (2) | (1) | (2) | (1) | (2) |
| | Largest amount received | Selection | Largest amount provided | Selection | Largest amount provided | Selection | Largest amount provided | Selection |
| Distance between residences (km) | 0.012<br>(0.001)*** | 0.006<br>(0.002)*** | | | 0.058<br>(0.229) | −0.017<br>(0.008) | 0.223<br>(0.198) | |
| *Purpose of largest transfer* | | | | | | | | |
| Farm investment | 24.476<br>(5.398)*** | | 23.330<br>(6.423)*** | | 4.569<br>(3.990) | | 4.174<br>(4.048) | |
| Farm inputs | 2.042<br>(3.714) | | 1.812<br>(4.331) | | 1.614<br>(1.730) | | 0.877<br>(1.764) | |
| Off-farm business expenses | 36.393<br>(9.828)*** | | 38.889<br>(11.776)*** | | 26.071<br>(3.946)*** | | 25.249<br>(4.004)*** | |
| School fees | 27.901<br>(4.094)*** | | 31.764<br>(4.785)*** | | 8.007<br>(1.834)*** | | 7.304<br>(1.876)*** | |
| Medical expenses | 4.223<br>(3.209) | | 2.268<br>(3.746) | | −0.344<br>(1.779) | | −1.071<br>(1.764) | |
| Ceremonial expenses | −3.109<br>(5.091) | | −6.551<br>(6.033) | | 2.634<br>(2.671) | | 2.520<br>(2.721) | |
| *Model summary* | | | | | | | | |
| Number of observations | | 2108 | | 10229 | | 2108 | | 10228 |
| Uncensored observations | | 484 | | 508 | | 437 | | 447 |
| *P* value Wald chi-squared | | 0.000 | | 0.000 | | 0.000 | | 0.000 |
| *P* value lambda | | 0.067 | | 0.081 | | 0.002 | | 0.000 |

Standard errors are given in parentheses.
*Significant at 10%; **significant at 5%; ***significant at 1%.

with others who are similar to themselves with respect to social status, human capital, wealth, etc. Social networks are not anonymous markets, and they are formed through an endogenous network formation process that has important implications for participants in them.

This finding is consistent with the findings of other studies. In the USA, the structure and composition of job-related networks have been shown to have career implications. Those who are not considered legitimate members of professional networks tend to be excluded from many opportunities. Burt (1998) found that the insider–outsider distinction in his study population was correlated with gender, which he suggested as an explanation to why upwardly mobile women tend to hit a 'glass ceiling' beyond which they are unlikely to be promoted. But the social capital of the successful is not always useful. Also in a study from the USA, Hurlbert *et al.* (2001) found that in the extreme situation of a natural disaster, the highly educated, despite having access to social capital, did not have access to the kinds of networks that facilitate reception of informal support.

This study finds that participation in transfer networks is dependent on one's resources. The poorest participate mainly in frequent but low-value transactions, mostly in kind. People of intermediate wealth engage more actively in transfers in kind than the poorer, but engage even less in cash transfers than the poor. But as people get wealthier, they engage more in cash transfers and less in transfers in kind. Wealth is associated with having a higher education, having off-farm earnings, and having a large landholding. Poverty is associated with no education and inability to recover from adverse shocks.

Women are more active in transfers in kind than men. Being credit-constrained in either the formal or informal credit market leads to a withdrawal from transfer networks, but having access to formal finance has only little effect on participation in transfer networks and virtually no effect on levels of transfers.

The most important criterion for achieving support through social networks is if it includes the right people. Two categories of network members stand out as particularly valuable: civil servants and military officers and people with off-farm businesses, who have higher and steadier cash earnings than the average subsistence farmer, and are good sources of cash transfers. It is correct to refer to transfer networks as kinship-based, because family members engage very frequently in transfers with each other, and they may be sources of large amounts of transfers, especially where a household receives regular remittances from a family member not resident in the household, often living quite far away.

The most common purpose of cash transfers is consumption, but when these villagers need to raise relatively large amounts of money, the purpose is often to pay school fees, and education of children clearly attracts large transfers. Bilateral transfers through networks are also valuable sources of finance for investments in income-generating activities. But there is no evidence in these data that social networks support households with seriously ill members, except for funeral expenses.

A primitive model of informal insurance depicts a rural village with identical households that engage in a village-wide risk and income-pooling arrangement. But despite the apparent universality of poverty in the communities studied here, it is clear that their populations are not uniform, and they do not form village-wide networks. Social networks form as individuals choose whom to interact with, and in what way. Strong selection mechanisms sort the population into participants and non-participants of transfer networks. Some degree of income and asset pooling can be assumed based on the observation that the studied social networks engage quite intensely in transfers in kind, dominated by transfers of goods and services that serve as farm inputs, such as borrowing farm implements, but also exchange purchased inputs for other goods or services. Transfers in kind are also important for consumption smoothing.

But there are clearly groups that are excluded, as they appear to be unable to participate in certain types of transfers. In particular the poorest appear to benefit less from these networks than those of intermediate or higher levels of wealth. Transfer networks provide the greatest benefits to wealthy households, and to those who have been able to include some key persons like civil servants in their networks, whose networks therefore consist of the most resourceful persons. There is a negative effect of age on participation in transfer networks, perhaps contrary to expectations.

This study cannot determine if the market for transfers within social networks is constrained on the supply or demand side, but some observations may provide clues for educated guesses. In a situation where the rural villages under study are being devastated by an AIDS pandemic that keeps people busy with funerals, easily observable during data collection, it is disturbing to find that bilateral transfers through social networks are not forthcoming when households suffer serious illness. What people manage to raise transfers for through their networks is school fees for their children and investments in farms and off-farm businesses. This may reflect that people want to channel their resources into investments for the future, and not squander them unproductively, but it leaves out purposes that are of vital importance to the affected households. Clearly, supply of transfers is a function of the suppliers' motivation to supply, and not just the costs the demanders face and their immediate needs.

These findings suggest a need for targeted intervention to help the socially excluded, and to fill the social security void where social networks fail. The excluded can be identified as the poorest, the uneducated and probably the elderly. Their needs for financial services include not only ways to finance investments in income-generating activities, but also access to appropriate savings instruments, which can help them self-insure. But no realistic measure of self-insurance can insulate households from the effects of sickness and death. Since social networks appear to be of limited help in those situations, policies are needed to improve provision of health care.

## Acknowledgements

This article was written for the World Bank under a short-term consultancy and is used here with their permission. The research it builds on was made possible with a grant from the Social Science Research Council, with funding from John D. and Catherine T. MacArthur Foundation, and grants from the Pew

Charitable Trusts, Mario Einaudi Center for International Studies and Cornell International Institute for Food, Agriculture, and Development. Partial funding was also provided by the US Agency for International Development through grant LAG-A-00-96-90016-00 to the Broadening Access and Strengthening Input Market Systems/Collaborative Research Support Program (BASIS CRSP), and the Strategies and Analysis for Growth and Access (SAGA) cooperative agreement, grant HFM-A-00-01-00132-00, as well as the Coupled Natural and Human Systems Program of the Biocomplexity Initiative of the National Science Foundation, grant BCS-0215890 (Pell). The International Centre for Research on Agroforestry (ICRAF) in Nairobi, Kenya Agricultural Research Institute (KARI) and Kenya Forestry Research Institute (KEFRI) also generously provided practical support during the fieldwork in Kenya.

## Appendix

**Table 10.A1.** Summary statistics for purposes of transfers.

| | Received | | Provided | |
|---|---|---|---|---|
| Cash transfers | Frequency | Percent | Frequency | Percent |
| Farm investments | 23 | 4.23 | 8 | 1.70 |
| Farm inputs | 63 | 11.58 | 68 | 14.47 |
| Off-farm business expenses | 7 | 1.29 | 11 | 2.34 |
| Residential building | 1 | 0.18 | 5 | 1.06 |
| School fees | 47 | 8.64 | 52 | 11.06 |
| Consumption | 296 | 54.41 | 232 | 49.36 |
| Medical expenses | 81 | 14.89 | 58 | 12.34 |
| Ceremonial expenses | 26 | 4.78 | 21 | 4.47 |
| Other or missing | | | 15 | 3.19 |
| Total | 544 | 100 | 470 | 100 |
| | Received | | Provided | |
| Transfers in kind | Frequency | Percent | Frequency | Percent |
| Farm investments | 12 | 1.50 | 15 | 1.84 |
| Farm inputs | 484 | 60.42 | 520 | 63.80 |
| Off-farm business expenses | 4 | 0.50 | 1 | 0.12 |
| Residential building | 3 | 0.37 | 2 | 0.25 |
| School fees | 3 | 0.37 | 5 | 0.61 |
| Consumption | 250 | 31.21 | 240 | 29.45 |
| Medical expenses | 18 | 2.25 | 14 | 1.72 |
| Ceremonial expenses | 25 | 3.12 | 18 | 2.21 |
| Other or missing | 2 | 0.24 | | |
| Total | 801 | 100 | 815 | 100 |

**Table 10.A2.** Summary statistics for binary variables (0 = No, 1 = Yes).

| Variable | No. Observations | Mean |
|---|---|---|
| *Respondent characteristics* | | |
| Site (0 = Vihiga, 1 = Embu) | 242 | 0.475 |
| Household head | 239 | 0.548 |
| Female | 242 | 0.579 |
| Household head has always lived here | 229 | 0.590 |
| Has not completed primary school | 239 | 0.159 |
| Has completed secondary school | 239 | 0.201 |
| Has completed technical school, college or more | 239 | 0.046 |
| Has house with thatched roof (= poor) | 236 | 0.000 |
| Has brick house (= rich) | 236 | 0.131 |
| Tea grower | 238 | 0.239 |
| Coffee grower | 238 | 0.458 |
| Raises cattle | 235 | 0.872 |
| Raises sheep or goats | 235 | 0.243 |
| Household receives remittances | 239 | 0.460 |
| Household head has a savings account in bank | 237 | 0.338 |
| Household head belongs to local savings institution | 237 | 0.329 |
| Credit constrained in formal credit market | 239 | 0.573 |
| Credit constrained in informal credit market | 242 | 0.029 |
| Household head has not recovered from earlier shock | 237 | 0.738 |
| Household has had a health shock last year | 242 | 0.161 |
| *Alter and reciprocal characteristics* | | |
| Retired | 2389 | 0.009 |
| Schoolteacher | 2389 | 0.047 |
| Priest, missionary or preacher | 2389 | 0.017 |
| Extension officer, vet, etc. | 2389 | 0.049 |
| Civil servant or military officer | 2389 | 0.056 |
| Petty trader, shopkeeper or business owner | 2389 | 0.096 |
| Village elder or chief | 2389 | 0.012 |
| Spokesperson or politician | 2389 | 0.008 |
| Same sex | 2389 | 0.681 |
| Close family member | 2383 | 0.219 |

**Table 10.A3.** Summary statistics for continuous and count variables.

| Variable | No. Observations | Mean | Standard deviation | Minimum | Maximum |
|---|---|---|---|---|---|
| *Transfers (dependent variables)* | | | | | |
| Number of times cash transfer received | 2389 | 0.697 | 2.468 | 0 | 60 |
| Largest cash amount received (US$) | 544 | 11.334 | 29.890 | 0.133 | 533.333 |
| Number of times cash transfer provided | 2389 | 0.548 | 1.883 | 0 | 40 |
| Largest cash amount provided (US$) | 470 | 7.379 | 15.086 | 0.080 | 133.333 |
| Number of times transfer in kind received | 2389 | 1.514 | 5.580 | 0 | 100 |
| Largest value received in kind (US$) | 801 | 3.465 | 24.836 | 0.133 | 666.667 |
| Number of times transfer in kind provided | 2389 | 1.702 | 6.100 | 0 | 100 |
| Largest value provided in kind (US$) | 815 | 2.284 | 8.756 | 0.013 | 200 |
| *Respondent and reciprocal characteristics* | | | | | |
| Household non-farm income per day (US$) | 239 | 1.183 | 1.987 | 0 | 15.518 |
| Household farm income per day (US$) | 239 | 2.070 | 2.738 | 0 | 21.322 |
| Per capita non-farm income per day (US$) | 239 | 0.244 | 0.415 | 0 | 3.653 |
| Per capita farm income per day (US$) | 239 | 0.437 | 0.529 | 0 | 3.046 |
| Respondent's age (years) | 231 | 51.952 | 15.431 | 14 | 88 |
| Area of cultivated land (acres) | 238 | 1.964 | 2.064 | 0.199 | 15 |
| Household labour days per month | 236 | 40.911 | 25.734 | 0 | 140 |
| Number of ROSCA memberships | 241 | 0.788 | 1.005 | 0 | 5 |
| Distance between ego's and alter's residence (km)[a] | 2373 | 2.847 | 23.362 | 0 | 350 |

[a]After excluding a single outlier living in America, recorded as a distance of 20,000 km. If this observation is included, the mean is 19.693 and the variance 580.76.

## References

Alderman, H., Chiappori, P.-A., Haddad, L., Hoddinott, J. and Kanbur, R. (1995) Unitary versus collective models of the household: is it time to shift the burden of proof? *The World Bank Research Observer* 10(1), 1–19.

Burt, R.S. (1998) The gender of social capital. *Rationality and Society* 10(1), 5–46.

Chabal, P. and Daloz, J.-P. (1999) *Africa Works: Disorder as Political Instrument*. Indiana University Press, Bloomington, Indianapolis.

Coleman, J.S. (1988) Social capital in the creation of human capital. *American Sociological Review* 94(Suppl.), S95–S120.

Heckman, J.J. (1979) Sample selection bias as a specification error. *Econometric* 47(1), 153–162.

Hurlbert, J.S., Beggs, J.J. and Haines, V.A. (2001) Social networks and social capital in extreme environments. In: Lin, N., Cook, K.S. and Burt R.S. (eds) *Social Capital: Theory and Research*. Aldine de Gruyter, New York.

Morduch, J. (2000) The microfinance schism. *World Development* 28(4), 617–629.

Platteau, J.-P. (1995) A framework for the analysis of evolving patron–client ties in agrarian economies. *World Development* 23(5), 767–786.

Udry, C. (1994) Risk and insurance in a rural credit market: an empirical investigation in northern Nigeria. *Review of Economic Studies* 61(3), 495–526.

# Index